D0806026

Energy and the unexpected

ENERGY AND THE UNEXPECTED

KEITH J. LAIDLER

Professor Emeritus of Chemistry
University of Ottawa

UNIVERSITY PRESS

OXFORD
UNIVERSITY PRESS

Great Clarendon Street, Oxford OX2 6DP

Oxford University Press is a department of the University of Oxford.
It furthers the University's objective of excellence in research, scholarship,
and education by publishing worldwide in

Oxford New York

Auckland Bangkok Buenos Aires Cape Town Chennai
Dar es Salaam Delhi Hong Kong Istanbul Karachi Kolkata
Kuala Lumpur Madrid Melbourne Mexico City Mumbai Nairobi
São Paulo Shanghai Taipei Tokyo Toronto

Oxford is a registered trade mark of Oxford University Press
in the UK and in certain other countries

Published in the United States
by Oxford University Press Inc., New York

© Keith J. Laidler, 2002

The moral rights of the authors have been asserted

Database right Oxford University Press (maker)

First published 2002

All rights reserved. No part of this publication may be reproduced,
stored in a retrieval system, or transmitted, in any form or by any means,
without the prior permission in writing of Oxford University Press,
or as expressly permitted by law, or under terms agreed with the appropriate
reprographics rights organization. Enquiries concerning reproduction
outside the scope of the above should be sent to the Rights Department,
Oxford University Press, at the address above

You must not circulate this book in any other binding or cover
and you must impose this same condition on any acquirer

British Library Cataloging in Publication Data

Data available

Library of Congress Cataloging in Publication Data

ISBN 0 19 852516 8

10 9 8 7 6 5 4 3 2 1

Typeset by Footnote Graphics,
Warminster, Wiltshire
Printed in Great Britain by
T.J. International Limited, Padstow, Cornwall

Preface

Not until the early nineteenth century was it recognized that energy is a distinct physical quantity, but it has proved to play a pivotal role in the advance of science. The history of energy is somewhat complex, and at times involved much controversy. It proved difficult to agree on exactly what energy is; heat in particular was slow to be recognized as a form of energy. The properties of energy also proved elusive. Restrictions as to how heat can be converted into work were elucidated only with difficulty and with much disagreement, sometimes acrimonious. That energy comes in packets, called quanta, was another subject of uncertainty. The fact that matter can be converted into energy was another topic of great significance that took some years to be appreciated.

The scientific investigations made with a view to elucidating the nature and characteristics of energy have had many important offshoots. One of them is our present understanding of the role of chance or probability in our world. This first became clear in the investigations made at the end of the nineteenth century concerning the conditions under which chemical and physical processes can occur. Previously determinism held sway among scientists; it was thought that one could in principle ascertain the exact state of the universe, and that all future events would inevitably follow. The work on the second law of thermodynamics cast considerable doubt on this conclusion. Even if one could know the exact state of the universe, chance would still enter into the course of events. Later the new quantum-mechanical ideas, based on energy considerations, put a somewhat different complexion on the whole problem. Now we know that it is absolutely impossible, being contrary to the laws of nature as we know them, even in principle to determine the present state of the universe. There is a fundamental uncertainty about the knowledge that can possibly be acquired; on those grounds alone prediction is impossible. We must always expect the unexpected.

Even that is not the full story. We now know, from modern chaos theory for which there is now overwhelming evidence, that there is a fundamental unpredictability about future events. We had all known that weather forecasting was unsatisfactory, but thought that it was just due to the complexity of climatic events. Chaos theory now tells us that there are more fundamental difficulties; the unexpected is inevitable, however carefully we could determine the details of the present situation.

This book is about these topics. It is written for the general reader who has no particular knowledge of any science but who is interested in understanding some important aspects of modern science. The approach is to some extent historical, partly because I think that some historical details make the subject more interesting. More important, I think that the history of science presented in an appropriate way does clarify important concepts, if only by showing how and why competent scientists have misunderstood them. It is encouraging, when struggling to understand an elusive topic like entropy, to know that a great scientist like Lord Kelvin did not understand it either, and that Einstein never accepted the uncertainty principle. I have not hesitated to make some simplifications, which will be obvious to my fellow scientists and which I hope they will forgive in view of the type of reader I have in mind; there is, after all, a wise maxim that says that a little inaccuracy can save a world of explanation. I have tried to keep to strictly benign inaccuracies.

Some scientific writers take the position that no equations should appear in a book intended for non-scientific readers, who will be irrevocably repelled by them. I think this is going too far. I am confident that simple equations, with adequate descriptions, make explanations easier rather than harder to understand. I think too that diagrams, inexplicably shunned even by some good scientific writers, can often help more than a multitude of words.

K. J. L.

December 2001

Acknowledgements

I am indebted to many people who have greatly helped me while I have been writing this book. Dr Alan Batten of the Herzberg Institute of Astrophysics, Dominion Astrophysical Observatory, has given me valuable advice on cosmological matters, and has read drafts of what I have written. For valuable discussions of chaos theory I am indebted to Professor Heshel Teitelbaum of the University of Ottawa. My many discussions with Professors John Holmes and Brian Conway of the University of Ottawa have always been helpful. I am indebted to Dr Walter Davidson, of the National Research Council of Canada, for help with my section about the latest work on neutrinos.

Dr June Lindsey and Dr George Lindsey, both physicists, have critically read the entire manuscript with great care and have made many constructive suggestions as to scientific content and style of presentation.

My son Jim Laidler has prepared most of the line drawings, and has been of great help with finding suitable portraits.

Since this book is primarily intended for readers who are not scientists, I needed the help of friends who would be willing to judge the book from that point of view, and give me appropriate advice. I am particularly grateful to Dr Christine Davenport, learned in history and the law but scantily schooled in science, for reading much of what I wrote in early drafts and advising me as to its suitability for the general reader.

Contents

A few points about mathematics

In this book I have tried to keep the mathematics to a minimum, and in particular have not used many mathematical equations. Some authors have eliminated equations completely from their books, and there is a popular mathematical theorem that sales of scientific books written for the public are halved for each equation included. I think that is going too far. Some mathematical equations, like Einstein's $E = mc^2$ and Boltzmann's $S = k \log W$, can greatly help the reader provided that they are properly explained. I have therefore not hesitated to use them and a few other equations.

Here are a few mathematical points to which some readers may need to be introduced.

Scientific notation

In this book we need from time to time to use very large or very small numbers, and it is useful to use what is called scientific notation to avoid writing out strings of numbers. Thus instead of writing 1,000,000, which is a million, we write 10^6, which is 10 multiplied by itself six times or one followed by six zeros. Similarly a billion, or 1,000,000,000, we write as 10^9. If we want to write 602,200,000,000, 000,000,000,000 (which happens to be the number of molecules in 18 grams of water) we write 6.022×10^{23}, which all will agree is a lot more convenient. The superscript numbers are called *exponents*. The 23 that appears in the expression is how many steps we have to take to move the decimal point from the right-hand end of the full expression in order to get 6.022.

A similar notation applies to very small numbers. Suppose we are dealing with 1/1,000,000 (one-millionth), which can also be written as 0.000001. To write it in scientific notation we note that one-millionth is $1/10^6$, and write it as 10^{-6}. A simple way of deciding on the exponent in such a case is to see how far we have to move the decimal point to the right in order to get a number between 1 and 9.99... . For 0.000001 we have to move it six times, and for 0.000234, for example, we have to move the decimal point by four steps to get 2.34, so that we write it as 2.34×10^{-4}. The value of the Boltzmann constant (which we will meet in Chapter 7) is 1.381×10^{-23}, and this means that it is 1.381 divided by one followed by 23 zeros.

An important point to note about numbers in scientific notation is that when the exponent changes by one, the number itself changes by a factor of 10. Thus

10^4 is ten times as large as 10^3, and 10^9 (a billion) is a thousand times as big as 10^6 (a million). The number 10^{-23} is a thousand times smaller than 10^{-20}.

Exponential quantities

Exponential quantities are used not just to represent numbers, but in another way. In Chapter 6 we will meet the quantity $\exp(-E/k_B T)$ which is just another way of writing

$$e^{-E/k_B T}.$$

By this we mean that the number e (= 2.71828...) is raised to the power of $-E/k_B T$. The number e, which has the value 2.71828..., often turns up in scientific work. When we write the exponential $e^{-E/k_B T}$ we mean that the number e is multiplied by itself $-E/k_B T$ times, just as when we write 10^2 we mean that two tens are multiplied together.

An important thing to note is that anything raised to the power of zero is one.

Logarithms

When we use logarithms we are doing the reverse of what we do in forming an exponential. We have seen that for 1,000,000 we can write 10^6. We can also say that the logarithm of 1,000,000, written as log 1,000,000, is 6. This kind of logarithm of a number is just the power to which 10 is raised to get that number.

This particular logarithm is said to be to the base 10, and is also called a common logarithm. The logarithm tables that were formerly used to make calculations, and have now been largely superseded by electronic calculators, employed common logarithms. When using common logarithms it is useful, to avoid ambiguity, to use the notation $\log_{10}$; thus $\log_{10} 1{,}000{,}000 = 6$.

In scientific work we more often use what we call natural logarithms, written as $\log_e$ or more often as ln. These are to the base e, which is the odd little number we used earlier, having the value 2.71828... . It is useful to note that natural logarithms are bigger than common logarithms by the factor 2.303. Thus

$$\log_{10} 1{,}000{,}000 = 6 \quad \text{and} \quad \ln 1{,}000{,}000 = 2.303 \times 6 = 13.818.$$

In using logarithms it is important to remember that a change in a logarithm means a larger change in the quantity measured. The pH scale, commonly used by gardeners and others, is a common logarithmic scale, and it is an inverse measure of acidity. Thus a solution of pH = 2 is 10 times as acidic as one of pH = 3 (10 times because it is a common logarithm, to the base 10). The Richter scale is also a common logarithmic scale, used for the amplitude of vibrations in an earthquake. In an earthquake measuring 7 on the Richter scale the average amplitudes are estimated to be 10 times as large as for one that is 6 on the Richter scale. The damage it causes will be much more than 10 times as great.

PREFIXES

The following prefixes are commonly used with the various units:

tera, T = 1 million million (10^{12})
giga, G = 1 billion = 1000 million (10^{9})
mega, M = 1 million (10^{6})
kilo, k = 1 thousand = 1000 (10^{3})
milli, m = one-thousandth (10^{-3})
micro, μ = one-millionth (10^{-6})
nano, n = one-billionth = one-thousand-millionth (10^{-9})
pico, p = one-million-millionth (10^{-12}).

ONE

Steam engines and artillery

Energy is Eternal Delight.

William Blake, *The Marriage of Heaven and Hell*, 1790

All of us today have a fairly good idea of what energy is, if only because we have become used to paying for it. We know that heat and work are forms of energy, and that heat can be converted into work and work into heat. Until about the middle of the nineteenth century, however, these topics were still a matter of confusion and dispute. Even the word 'energy' is a comparatively new one. It does not appear in the Bible, and Shakespeare never used it. The poet Alexander Pope made a passing reference to 'energy divine'. But Isaac Newton, who did such pioneering work in mechanics, never used the word—or if he did he did not make much of it—and it was rarely used until the nineteenth century.

Reaching the correct conclusion about the nature of heat and its relationship to mechanical work was by no means a simple matter. Much help came from purely technical and empirical work on the development of steam engines. The steam engine was invented, and indeed brought to a high degree of perfection, by men who had no training in science and little knowledge of it. Only after a particularly efficient steam engine had been built, by James Watt in the latter part of the eighteenth century, did scientists begin to investigate how it worked. Their conclusions were embodied in the two basic laws of thermodynamics. These laws, so important in all of science today, thus owe their origin not to people who were seeking the truth for its own sake, but to the empirical efforts of a few extremely ingenious engineers.

The first steam engine of much practical value was invented by Thomas Newcomen (1663–1729), who was born in Dartmouth, England, and began his career as a blacksmith. By 1712 he had constructed a steam engine for use mainly in pumping water out of coal mines (Fig. 1). Its principle was simple. It had a single cylinder with a piston connected to a pivoted wooden beam. Steam from a boiler was admitted to the cylinder and the piston rose; then cold water was admitted to the cylinder, causing the steam to condense and the piston to fall. The pivoted beam performed about 12 strokes per minute, and a system of valves automatically controlled the admission of steam and of cold water to the cylinder.

If Archimedes (*c.* 287–212 BC) had been shown a Newcomen engine he would have had no difficulty in understanding how it worked. In a steam engine (which was at first called a 'fire engine'), the steam admitted into the cylinder moves the piston and performs mechanical work. After the steam is cooled, so that it con-

Fig. 1 An engraving showing an early steam engine at Griff, Warwickshire. (From an engraving in J. T. Desagulier's *A Course in Experimental Philosophy*, 1744 edition.)

denses to liquid water, much less work is required to move the piston back to its original position, and then the cycle can continue. A net amount of work is therefore performed by the engine. We know now that when this work is performed there is inevitably a loss of some of the heat provided to the engine, but this was by no means clear until well into the nineteenth century, as we will see later in this book.

Newcomen engines generated about 5 horsepower, the horsepower being a unit later invented by James Watt in order to express the power of an engine in terms of the number of horses it replaced. We can also make an estimate of what is now called *thermodynamic efficiency*; this is the ratio of work actually performed by an engine to the maximum amount of work that would be performed if all of the heat supplied by the boiler could be completely converted into work. For a Newcomen engine this efficiency is estimated to be about 1 per cent. In view of this low value it may seem surprising that a few Newcomen engines were still in practical use at least until the third decade of the twentieth century. They were at coal mines, where there is always a residue of poor-quality coal which cannot be sold; it was better to burn it in an inefficient engine than to throw it away.

The great name in connection with steam engines is James Watt (1736–1819; Fig. 2). The circumstances in which Watt invented his much more efficient steam

Fig. 2 James Watt (1736–1819), who is famous for his invention of the separate condenser which led to his great improvements in the design of steam engines. As a unit of power (rate of doing work) he introduced the 'horsepower', but the modern unit is the watt (symbol, W), named in his honour (1 HP = 745.7 W). Watt was a man of fine character, and British prime minister Lord Liverpool said of him: 'A more excellent and amiable man in all the relations of life I believe never existed.'

engine are especially interesting, and remind us that chance as well as mechanical genius is always involved in an important technical innovation. Watt was born in Greenock, Scotland, and in 1755, at the age of 19, he went to London to learn the trade of instrument maker. On his return to Glasgow a year later he wanted to set up a business as an instrument maker, but the Hammersmen's Guild put difficulties in his way since he had never served an apprenticeship. The University of Glasgow, however, allowed him to practise his trade on the university premises, and there he soon established a reputation for ingenuity and persistence. At the time the university was remarkably rich in talent, the economist and philosopher Adam Smith (1723–1790) and the chemist Joseph Black (1728–1799), being two of its professors. During his stay in Glasgow, until 1774, Watt became friendly with some of the professors and students, notably Joseph Black who was doing work of great importance on the subject of heat. Watt also became friendly with John Robison (1739–1805), Black's student and successor as professor of chemistry. Undoubtedly these relationships helped Watt to attack his technical problems in a very scientific way.

At the time the professor of natural philosophy (physics) at Glasgow was John Anderson (1726–1796), a rather remarkable man. He had previously been professor of oriental languages, and he had a strong social conscience. He believed that his physics lectures should be available to artisans and others who were not members of the university, and he threw his classes open to them. These 'anti-toga' classes did not receive the approval of the university authorities, and somewhat hard feelings developed. As a result Anderson left all of his considerable fortune to found a rival institution in Glasgow. It became known as Anderson College or the Andersonian Institution, and evolved into the present University of Strathclyde.

For demonstrations in his physics classes, Anderson had a model of a Newcomen engine. It worked badly, performing only a few strokes before stalling, and after finding that instrument makers in London were unable to help, Anderson asked Watt to overhaul it. Watt went about this task with great persistence, and carried out a number of investigations on the thermal effects of mixing steam and water. He concluded that the main trouble with the model was that, because the surface-to-volume ratio was much larger than in the full-scale engines, the loss of heat was relatively much greater. ('Scaling' effects of this kind had previously been recognized by Newton.) We now know that there were other factors besides the one mentioned by Watt. The cylinder of the model was made of brass, while in the engines themselves they were of iron and would be coated internally with iron oxide; this is another reason for greater heat loss in the model. In addition, the wall thickness in the model was disproportionately large, so that the cylinder had a larger heat capacity. Watt was able to get the model working (but only just!) by careful control of the amount of cooling water added, and by reducing leakage at the piston.

While working on the model of the Newcomen engine, Watt realized that it had a fundamental flaw. Heating up the cylinder with steam, and then cooling it with water, obviously involves much unnecessary wastage of heat and loss of efficiency. Watt then had the most important of his many innovative ideas, the *separate condenser*. We even know just when and where Watt had this idea: it was on Easter Sunday 1765, when he was walking past the Golf House on Glasgow Green. His suggestion was that there should be two cylinders connected together, one always kept hot, and the other, the condensing cylinder, always cold (see Fig. 3). At first the compression of the hot cylinder was brought about by the atmospheric pressure, but in 1769 Watt realized that this produced unnecessary cooling, and that it was better to let steam do the compression; this was his second important innovation. In the same year he introduced his so-called *expansive principle*; instead of continuously admitting steam to bring about the compression, he cut off the supply and let the pressure fall. He estimated that greater efficiency resulted in this way.

In order to put his ideas into practice Watt had to get some extensive financial backing. After some abortive enterprises he entered into a partnership with Matthew Boulton (1728–1809), a highly successful and prosperous manufacturer in Soho, near Birmingham. Watt had obtained a patent for his engine in 1769, and manufacture was commenced at the Soho Engineering Works in 1774. Altogether about 500 engines were built during the Boulton–Watt partnership. The thermodynamic efficiencies of the first ones were about 8 per cent, but by the end an efficiency of about 19 per cent had been achieved. This was quite creditable since, as we will see in Chapter 4, the second law of thermodynamics imposes a *maxi-*

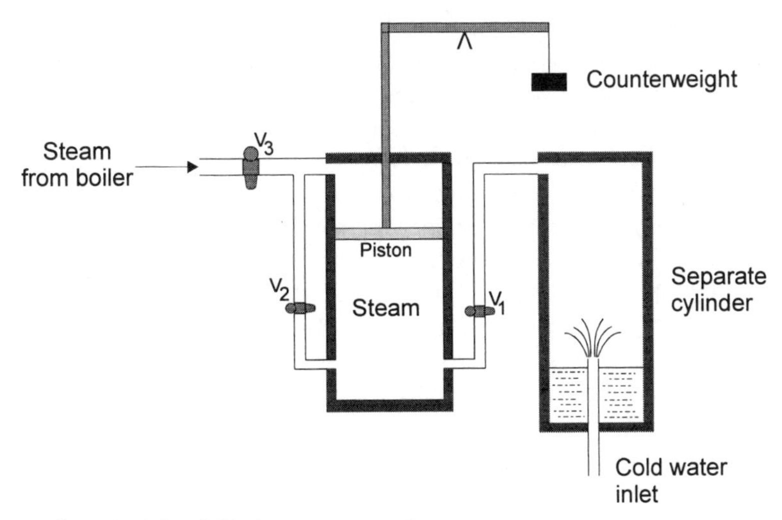

Fig. 3 The principle of Watt's separate condenser.

mum possible efficiency of about 25 per cent under the conditions of the Watt engines.

One reason for the great success of the Boulton–Watt partnership was the very different and complementary temperaments of the two men. Watt, though a kindly man of fine character, was a somewhat dour Scot who tended to take a pessimistic view of things. Boulton was a large, cheerful man who was always optimistic, and he gave Watt much encouragement. On one occasion when Watt was unduly depressed, Boulton wrote to him suggesting that he should say twice daily the 'Scotch prayer', which reads 'The Lord grant us a guid conceit of airselves'. It should be explained that at the time the word 'conceit' did not have its present pejorative meaning, but instead meant a justifiable self-esteem. Watt was certainly entitled to much self-esteem, but was too modest to realize it.

Another reason for success was that Watt and Boulton collaborated with John Wilkinson (1728–1808), a highly inventive ironmaster. Wilkinson's firm was able to bore cylinders much more accurately than had hitherto been achieved, and this greatly improved the efficiency of the steam engines. Another of Wilkinson's achievements was that he was the first to float an iron ship. Its launching in 1787 created a great sensation: it was thought that material denser than water would never float, and ships had always been made of wood. Wilkinson constructed for himself an iron coffin, but his foresight in so doing had its limitations: by the time of his death his cross-sectional area had increased so much that he could not be squeezed into it.

Watt's engines found many applications. The first locomotive engine was built in 1784 by William Murdoch (1754–1839), who was Watt's assistant. At the time he was building stationary engines in Cornwall, and while living in Redruth built a model locomotive. He decided to make use of the steep banks of a lane leading to the local church. When he made his first trial, at night, the engine moved so fast that it out-distanced him and almost ran down the rector of the church who happened to be walking along the road. As he ran away in terror the rector assumed the fiery monster to be the embodiment of the devil.

However, little use was made of the Watt engines in locomotives. The reason was that Watt was strongly opposed to the use of high-pressure steam, for reasons of safety. With improvements in engineering techniques, however, it became possible to construct high-pressure steam engines that were quite safe. The first successful high-pressure steam carriage was made by Richard Trevithick (1771–1833) in 1801. During the next few years Trevithick designed a number of stationary steam engines, steam carriages, and steam locomotives. He constructed the first steam locomotive that travelled on tracks in 1804, and showed that it provided considerable traction. Nothing that he did, however, was commercially successful, and he died in debt.

With high-pressure steam it was usual not to use a condenser, but simply to vent the steam to the surroundings. The non-condensing high-pressure engines were less efficient than the Watt engines (having a thermodynamic efficiency of about 10 per cent rather than 19 per cent), but the absence of a condenser made them much more compact. With a locomotive, as opposed to a ship, compactness was of great importance, and all the major developments of steam locomotives involved the use of non-condensing engines.

Without in any way detracting from the genius of Watt, we may reflect on the amount of luck that was also involved in his work on the steam engine. Several circumstances fitted together very neatly. Watt had just established himself at the University of Glasgow, with the right skills, when John Anderson had trouble with his model Newcomen engine and asked for his help—and Watt was perhaps the only person in the world competent to give it. Watt, needing financial support for his invention, began an association with Matthew Boulton, one of the few men in the world capable of leading it to industrial success. And just at the right time they gained the cooperation of John Wilkinson, who could bore a cylinder more precisely than anyone else in the world.

Perhaps the most important factor leading to the success of Watt's work on the steam engine was that besides being a great engineer he was also, without training but by instinct and through the influence of his associates, an excellent scientist. His technical developments were always supported by his work in pure science. He carried out fundamental experiments on the condensation of steam and, quite independently of Joseph Black, discovered *latent heat*—the heat that is required, for example, to convert water into steam, and is released when the steam condenses. Watt was, incidentally, perhaps the first person to realize that water is a chemical compound and not an element.

Over the centuries there has been much confusion and disagreement about the nature of heat. Today we know that it is a form of energy, arising from the movement of the atoms and molecules of which matter is composed. In hydrogen gas at ordinary temperatures the molecules are moving at high speeds, and we say that they possess kinetic energy of translation, the word translation referring to the fact that they move from one place to another. Each molecule of hydrogen consists of two hydrogen atoms connected together by a chemical bond which acts quite like a spring, and chemists use the symbol H_2 to represent hydrogen. The molecules are in a constant state of vibration, and in addition they rotate. The energy of a gas exists therefore in three forms: translational, vibrational, and rotational. If we raise the temperature of the gas we give it more energy, which becomes distributed between the translational, vibrational, and rotational modes of motion of the molecules. Heat is therefore a manifestation of the energy contained in a substance, and is in no way a substance itself.

That is the modern view of heat, and it is interesting that it is the view held in the early seventeenth century by Francis Bacon (1561–1626), who although primarily a lawyer and statesman had thought much about science and technology. From his study of the evidence he produced strong arguments for believing that heat is not a substance but a form of energy: in his words, 'heat itself, its essence and quiddity [nature], is Motion and nothing else'. This was also the view of the great investigators Robert Boyle (1627–1691) and Isaac Newton (1642–1727). Both of them regarded matter as composed of atoms, and their idea was expressed very clearly by John Locke (1632–1704): 'Heat is a very brisk agitation of the insensible parts of the object [i.e. of the atoms], which produces in us that sensation from which we denominate the object hot; so that what in our sensation is heat, in the object is nothing but motion.' Locke is now chiefly remembered as a philosopher, but he was active as a physician, and did a certain amount of work in experimental chemistry.

In the eighteenth century, however, this correct view of heat was seriously challenged by people who concluded that heat is a substance. One investigator who came to this conclusion, and who because of his prestige exerted a strong influence, was the great French chemist Antoine Lavoisier (1743–1794). He carried out some investigations that convinced him that heat was a substance of a rather special kind, in that it had no weight. Lavoisier referred to it as an 'imponderable fluid', and even included '*calorique*' in his list of the chemical elements. Since most of Lavoisier's other scientific ideas were certainly correct his idea that heat is a substance was taken seriously by many scientists of his time.

The experiments of the Scottish chemist Joseph Black, at the University of Glasgow and later at the University of Edinburgh, also seemed to lead to the conclusion that heat is a substance. Black first clearly distinguished between temperature and heat, and showed how temperature measurements can be used to determine the quantity of heat. Black was struck by the fact that mercury, which is very dense, has a lower heat capacity than an equal volume of water. By this we mean that to raise the temperature by a given amount it does not take as much heat for mercury as for water. He thought, not unreasonably at the time, that more internal motion should be possible in a material of greater density; therefore, he concluded, heat cannot be motion. He was wrong, and this is an instructive example of how the most careful experiments and intelligent reasoning can lead one astray if the time is not ripe.

Watt obviously had an excellent understanding of the operation of steam engines, but he never concerned himself much with just what heat is. On the whole he tended to accept the view of his friend Joseph Black that heat is a substance. Nothing in the operation of a steam engine points strongly towards one particular explanation of heat. Heat can be a form of motion which in an engine is converted into the movement of pistons; this we now know to be correct. Alternatively, the operation of an engine may be compared with a waterfall causing a wheel to turn (Fig. 4). It can be supposed that the heat is a substance that flows from a higher to a lower temperature, just as water flows from a higher to a lower level. When water causes a water wheel to turn there is no loss of water, the wheel being turned by the force of the water. In the same way, the force of falling heat could cause the piston in an engine to move and perform work, without any loss of heat. It was not for several decades that this second explanation was completely discarded, and as we shall see in Chapter 3 the theory of the steam engine greatly helped to settle the matter.

Strong evidence for the correct explanation, that heat is a form of motion, was presented in the late eighteenth century by experiments on the boring of cannon. These were carried out by the remarkable administrator and investigator Benjamin Thompson (1753–1814; Fig. 5) who is usually known today by the name Count Rumford. Born in Woburn, Massachusetts, of parents who kept a small farm,

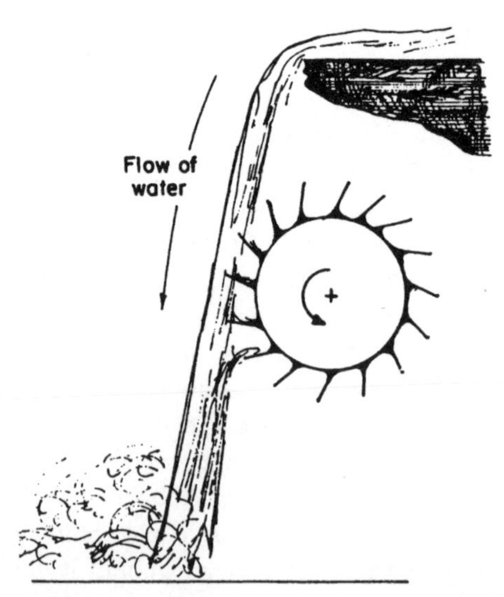

Fig. 4 Falling water can perform work by turning a water wheel, and it was thought that a steam engine operates by the action of heat falling from a higher to a lower temperature.

Fig. 5 Benjamin Thompson (1753–1814), American-born scientist and inventor, better known as Count Rumford. He had a colourful career in England and Bavaria.

Thompson received an adequate education, augmenting his schooling by avid reading. At the age of 14 he was apprenticed to a storekeeper, but the work was not to his liking and he soon abandoned it. Later he boarded with wealthy families, acting as a tutor to their children. By the age of 18 he was tall and handsome, with carefully powdered hair, piercing blue eyes, and an engaging manner. He caught the eye of a wealthy widow, Mrs Pierce, who was 14 years his senior, and within four months of their first meeting they married in November 1771. They settled at his wife's home in Concord, New Hampshire, which had formerly been called Rumford.

This marriage at once converted the impecunious young tutor of modest family background into a wealthy landed gentleman, and he took every advantage of his improved circumstances. Throughout his life he had a sharp eye for the main chance, and within a short time the Governor of New Hampshire had granted him a major's commission in the 2nd Provincial Regiment, a meteoric rise for a young man of no military experience. Relations with the mother country were at the time strained, with about two-thirds of the colonists favouring the cause of independence. Thompson favoured the British side, which caused him some difficulties, and in 1776, on the outbreak of the War of Independence, he fled to England, abandoning his wife and baby daughter. His wife continued to live in America for some years but he never saw her again and had almost no contact with her. His daughter joined him after her mother's death, but their relationship was always an uneasy one.

In England he gave valuable information to the government on the situation in America (in other words, he acted as a spy) and received an important appointment in the Colonial Office. During his stay in England he became acquainted with many prominent people including King George III, who at first was friendly with him but like most people soon became disillusioned. He had his portrait painted by the great Thomas Gainsborough in full military regalia. He carried out some experiments with gunpowder, and as a result was elected a Fellow of the Royal Society in 1781, at the early age of 27. In the following year he returned to America for a time with the rank of lieutenant-colonel in the Regiment of Artillery, again a surprising promotion for one with no military experience. He was later promoted to full colonel and soon retired with a lifetime pension.

In 1784, finding himself with not enough to do in England, Thompson decided to enter the service of the Elector of Bavaria, with whom in his characteristic style he had become acquainted. To work in Bavaria he had to obtain the permission of King George from whom he already held a commission; the king granted permission with an alacrity that might be thought suspicious, and at the same time knighted him, perhaps partly to speed him on his way but also to improve the somewhat strained relationships between Britain and Bavaria. There he again experienced a meteoric rise, remarkable in one who spoke no German and little French, and who was a Protestant in a strongly Catholic country.

At the time, Bavaria was in a sad condition, and reform was badly needed. There was much poverty: in Munich the proportion of homeless beggars was about 5 per cent, and there was much violent crime. In a short time Colonel Sir Benjamin Thompson brought about a number of remarkable improvements. He reformed the army, established a military academy, planned a poor-law system, spread the knowledge of nutrition and domestic economy, and improved the breeds of horses and cattle. He arranged for the conversion of a prison into a House of Industry (workhouse) into which the beggars were rounded up and required to work at making military uniforms. It began its operation on New Year's Day 1790, and was remarkably successful. The former beggars were well cared for, being given good food, some education and other training, and adequate medical care. Another of his accomplishments in Munich was the establishment of what is now called, because of his connection with it, the English Garden (although American born, his repudiation of his country and support of the British had, of course, made him an Englishman). This garden was built in the style of Kew Gardens in London, having paths, bridges, lakes, a Chinese pagoda, a concert hall, and an amphitheatre. It was opened to the public in 1791, and in its time was the finest park in Europe.

For these services Thompson was made a major-general in the Bavarian Army, Minister of War, Royal Chamberlain, a privy counsellor, and chief of police (the *Mikado*'s Pooh-Bah did not do any better). He was also created a Count of the Holy Roman Empire, choosing the title Count Rumford, the former name of the town of Concord where he had lived. He spent 1795–1796 in England, and endowed two Rumford medals of the Royal Society for research in light and heat.

At about he same time, perhaps to show that there was no ill-feeling at any rate on his side, he endowed two medals for the American Academy of Arts and Sciences (this academy, founded in Boston in 1780, is not to be confused with the National Academy of Sciences, based in Washington). These awards caused some awkwardness since he insisted on being the first winner of each of the prizes. On his return to Munich in 1796 he found that the king, threatened by both France and Austria, had fled, and he was made lieutenant-general and commander-in-chief of the Bavarian Army and president of the Council of Regency, responsible for the defence of Munich, which in the end he achieved without a drop of blood being spilt.

He decided to upgrade the artillery and in particular to arrange for the manufacture of heavy brass cannon. He personally supervised the boring of the cannon barrels. Each was cast in the form of a solid cylinder and was bored by being attached horizontally to a lathe. The drill bit, made of hardened steel, was held stationary. A revolving shaft attached to the rear of the barrel caused the barrel to turn on its axis at a rate of about 32 revolutions per minute. The power for turning the barrel was provided by two horses, a system of gears transmitting the motion to the shaft of the lathe.

Rumford was soon impressed by the fact that much frictional heat was continuously produced, and he made measurements of the amounts of heat. For this purpose he arranged for the casting of a specially shaped cannon barrel that could be insulated against loss of heat, and replaced the sharp boring tool with a dull drill to increase the friction and thus generate more heat. By immersing the drill in a tank full of water and making temperature measurements he was able to determine the amounts of heat produced. In his later report to the Royal Society he wrote that he 'perceived, by putting my hand into the water and touching the outside of the cylinder, that Heat was generated: and it was not long before the water which surrounded the cylinder began to be sensibly warm'.

He was particularly impressed by the fact that, as he said in his Royal Society paper, 'at 2 hours and 30 minutes it ACTUALLY BOILED. It would be difficult to describe the surprise and astonishment expressed in the countenances of the bystanders, on seeing so large a quantity of cold water heated, and actually made to boil, without any fire.' He continued, in a personal style that unhappily no editor today would allow in a scientific paper: 'Though there was, in fact, nothing that could justly be considered as surprising in this event, yet I acknowledge fairly that it afforded me a degree of childish pleasure, which, were I ambitious of the reputation of a *grave philosopher*, I ought most certainly rather to hide than to discover.'

Rumford established that no weight change occurred during the process, and that the metal shavings had the same properties as the unbored metal. He was particularly impressed by the lavishness with which the heat was produced: 'We have seen that a very considerable quantity of Heat may be excited in the friction of two metallic surfaces and given off ... without interruption or intermission, and without any sign of diminution or exhaustion.' From these investigations he concluded that heat could not be a substance, a conclusion he expressed clearly and convincingly: 'anything which any *insulated* body ... can continue to furnish *without limitation*, cannot possibly be a *material substance*.' Heat, he concluded, cannot be explained 'except it be Motion.' In his cannon-boring experiments there was thus a conversion of mechanical work into heat, and he obtained a value for the 'mechanical equivalent of heat', which is the amount of work that is required to produce a given amount of heat.

Although the results of these experiments were impressive, they by no means convinced many scientists that heat is a form of motion rather than a substance. The scepticism with which his work was received may have been largely due to his personal unpopularity, arising from his general behaviour which most people found insufferable. Later appraisals of his heat experiments, by people who did not know him personally, have been kinder. The Irishman John Tyndall (1820–1893), for example, who succeeded Michael Faraday as director of the Royal Institution in 1854, wrote in 1871 of Rumford's work: 'Hardly anything more powerful against the materiality of heat has been adduced, hardly anything conclusive in the way of establishing that heat is what Rumford considered it to be, Motion.'

In 1799 Rumford left the Bavarian service and returned to England, where he became associated with the Society for Bettering the Condition and Increasing the Comforts of the Poor. He proposed the founding of a public institution for the diffusion of knowledge, for giving public lectures, and for applying science to practical ends. This suggestion led to the opening in London, on 7 March 1799, of the Royal Institution of Great Britain, which continues to this day to do outstanding work in a variety of fields. In 1801 Rumford persuaded Humphry Davy (1778–1829), still only 22 years old, to join the Royal Institution as assistant lecturer in chemistry. This was a happy choice; a year later Davy had been promoted to become professor of chemistry and director of the laboratories, and he went on to do research of great distinction, being knighted in 1812.

Rumford's association with the Royal Institution was short-lived, as he soon quarrelled with its managers. In 1804 he moved to Paris where he renewed his acquaintance with Lavoisier's wealthy widow. Born Marie-Anne-Pierette Paulze, she was brought up in a cultured household and married Lavoisier, sometimes known as 'the father of modern chemistry', when she was only 13 and he was 28. Because Lavoisier had become a 'farmer-general' (collector) of taxes he was unpopular with the French revolutionaries and was guillotined in 1794. Madame Lavoisier first met Rumford in November 1801, and later became his mistress, marrying him in October 1805; he was 52 and had been separated from his now-deceased wife for 30 years, she was 45 and a widow for 11 years. They were two strong-willed people, used to having their own way, and were on friendly terms when they were not living together in a state of matrimony, but quarrelled incessantly when they were. The fact that the bride insisted on being known after marriage as Countess Lavoisier de Rumford was, in those days, a bad sign. Within a couple of months Rumford wrote to his daughter complaining about his wife, and after a year he was describing her as a 'female dragon'. They soon separated, and after petty squabbles about financial matters the separation became official in June 1809.

Count Rumford died, suddenly and unexpectedly, on 21 August 1814 at the age of 61. The funeral at Auteuil was attended by only a few people, neither his daughter nor his ex-wife being present. The inscriptions on his grave in the cemetery at Auteuil are now illegible. The main beneficiary of his will was, rather surprisingly, Harvard University, which to this day has a Rumford professorship devoted to lectures concerned with the 'utility of the physical and mathematical sciences for the improvement of the useful arts'.

Rumford's scientific and organizational achievements were outstanding. One of his inventions, that of the so-called 'Rumford fireplace', brought him considerable wealth. It warmed much more effectively than a conventional grate, and greatly reduced the escape of smoke into a room; modern fireplaces are based on the principles he established. In connection with the workhouse in Munich he designed very effective lamps and kitchen ranges that could be used for feeding large numbers of people. In his work with the Bavarian Army he made useful experiments on the effectiveness with which various types of clothing kept people warm, and his findings were of considerable practical importance.

Another of his scientific achievements is now largely forgotten. In 1794 Elizabeth Fulhame, about whose personal life we know practically nothing, not even her years of birth and death, published a remarkable book, *An Essay on Combustion*. This included an account of some interesting experiments on photo-imaging, involving the creation of images on cloths impregnated with gold and silver salts. This work was done more than four decades before the technique of photography was introduced in 1839, and perhaps because it was so far ahead of its time her experiments attracted little notice. One who did pay attention to them was Rumford, who carried out experiments along the same lines as those of 'the ingenious and lively Mrs Fulhame', a description that suggests that he knew her personally. He reported to the Royal Society in 1798 that his experiments, inspired by hers, had completely confirmed her conclusions.

Aside from his outstanding ability Rumford must have had considerable personal charm, but of a superficial kind. Throughout his career he had many mistresses, most of them socially prominent women, including some Bavarian countesses, perhaps the Electress of Bavaria, and Lady Palmerston, the mother of the future British prime minister. He had several illegitimate children. His personal relationships usually soured rapidly, since his treatment of people was complex and inconsistent. On the one hand he loved humanity in the abstract, and put tremendous effort into easing the lot of the disadvantaged. On the other hand his behaviour towards those with whom he came into personal contact was arrogant and overbearing. He was admirably and pithily described by William H. Brock in an article in 1980 as 'a loyalist, traitor, spy, cryptographer, opportunist, womanizer, philanthropist, egotistical bore, soldier of fortune, military and technical adviser, inventor, plagiarist, expert on heat (especially fireplaces and ovens) and founder of the world's greatest showplace for the popularization of science, the

Royal Institution'. His own assessment of himself, in a letter to his former patron Lord Sackville, is uniquely and unconsciously revealing:

> No man supported a better moral character than I do, and no man is better satisfied with himself.

Two

Red blood and electric motors

Question your desires;
Know of your youth, examine well your blood.
William Shakespeare, *A Midsummer Night's Dream*

It was not for half a century after Rumford's work on the boring of cannon that the majority of scientists became convinced that heat is a form of motion. Two men played a particularly important role in bringing about the conversion of the unbelievers: the German physician Julius Robert Mayer and the English brewer and amateur scientist James Prescott Joule.

Julius Robert Mayer (1814–1878; Fig. 6) was born in the year of Rumford's death, in the south German town of Heilbronn. At the age of 10 he thought about perpetual motion machines, but decided not to go so far as to try to create

Fig. 6 Julius Robert Mayer (1814–1878), who in 1842 produced convincing arguments, based on metabolic processes, in favour of the principle of conservation of energy.

energy out of nothing. Instead he attempted to convert a small amount of energy into a larger amount. He arranged for a small water wheel to operate in the town brook, and by means of gears tried to amplify its power, with the idea of making it drive heavy machines. We are not surprised that he failed, but the experiment was helpful to him in his later understanding of mechanical work.

At university he studied medicine, but little physics, which was a serious disadvantage to him in his later investigations. He passed his state medical examinations in 1840 and decided, against his father's advice, to take a sea voyage in order to explore the world. He signed on as a doctor on the Dutch merchant ship *Java*, which sailed out of Rotterdam bound for the Dutch East Indies. When the ship was off Indonesia some of the crew succumbed to an epidemic, and Mayer performed a number of blood-lettings from veins in the arm. In those days bleeding was still an accepted form of medical treatment, and was particularly common in the tropics since it was felt that relieving even perfectly healthy people of some of their blood would counteract the effects of the heat. By that time it had been established by Antoine Lavoisier that warm-blooded animals are kept warm by the combustion of food which is oxidized by oxygen from the air. On this basis it could be understood why the blood in the arteries, which carry it from the lungs, is rich in oxygen and therefore bright red. The veins, on the other hand, are transporting the blood back to the lungs, and the blood in them is therefore depleted in oxygen and has a dark purple–red colour. This being so, surgeons who performed blood-lettings always took it from a vein and not from an artery; bleeding from an artery is harder to stop than bleeding from a vein.

Mayer found to his surprise that when he removed blood from a vein it was unusually red, as if it had come from an artery instead of a vein. To explain his observation that venous blood is significantly redder in the tropics than in a colder climate he suggested that in warmer weather there is a lower metabolic rate, less oxidation being required to keep the body warm. As a result of the smaller consumption of oxygen there is less change in the colour of the blood and less contrast in colour between venous and arterial blood. This rather simple observation, which many others must have noticed without paying attention to it, caused Mayer to think deeply about the relationship between food consumption, heat production, and work done. He also made a careful study of some of the heat studies that had been made by various investigators, but did few experiments on the subject himself.

On the basis of evidence and arguments of this kind, Mayer arrived at the conclusion that heat and work must be interconvertible—the heat from the Sun, muscular exertion, the energy from the metabolism of food must all be forms of the same physical entity. He considered all of them to be different forms of what he called force, by which he meant energy. Mayer was a deeply religious man, and he caused much trouble for himself by allowing metaphysics to enter into his thinking. He was convinced by the work of Lavoisier that matter is indestructible, and his own observations led him to believe that energy was also indestructible. In this he was on solid ground, but he unwisely introduced a third indestructible entity, the human soul.

On his return home in early 1841 Mayer prepared a scientific paper on his ideas; it was couched in metaphysical terms, and since he was ignorant of mechanics he made many elementary errors. He also used personal jargon that was inappropriate for a scientific paper. In June 1841 he submitted it to the leading German journal of physics, the *Annalen der Physik und Chemie*. At the time the editor of the *Annalen* was the distinguished physicist Johann Christian Poggendorff (1796–1877) who was editor from 1824 until his death 53 years later; it became so closely associated with him that it was commonly known as *Poggendorff's Annalen*. Unimpressed by the metaphysics and faulty physics Poggendorff rejected the paper out of hand. Mayer was angry and frustrated at this treatment, but it had the effect of stinging him into action. He soon agreed that the paper did have serious limitations and revised and improved it extensively, making it much more convincing. Nine months after submitting the original paper he submitted his revised paper, this time not to Poggendorff but to the *Annalen der Chemie und Pharmacie* whose editor was the distinguished German chemist Justus von Liebig (1803–1873). It was at once accepted and it appeared on 31 May 1842, which delighted Mayer since it was his wedding day.

Mayer later claimed that this paper established his priority for the principle of conservation of energy; by that time Rumford's work had been largely forgotten. Even in its revised form, however, the paper was not an impressive one, as it relied a good deal on rather casual observations and intuitive impressions, and not at all on the results of carefully planned experiments. Over the next few years Mayer published further articles on the subject, paying particular attention to the physiological aspects of the problem. In a book that appeared in 1845 he was critical of Liebig's ideas about animal metabolism, which involved a vital force which Mayer rightly considered to be unsatisfactory. However, since he did not belong to the scientific élite of the time his ideas were at first either ignored or ridiculed. He was a proud and sensitive man and this treatment had a serious effect on his mental stability. He also experienced some personal misfortunes: within a period of three years one of his sons and two of his daughters became ill and died before they reached the age of 3. In 1850, overcome by the strain, he attempted suicide by jumping out of a third-flood window and falling nearly 10 metres. He survived, but suffered severe injuries which caused a further deterioration in his mental condition. Several times he was confined in mental institutions, and sometimes had to be restrained in a strait-jacket.

His life, however, had a happier ending, since there was final recognition of his accomplishments. He was elected a corresponding member of the French Académie des Sciences, received some honorary degrees, and in 1871 received the Copley Medal of the Royal Society. He died of tuberculosis in 1878 at the age of 63.

The work done by James Prescott Joule (1818–1889; Fig. 7) was of a very different character from that of Mayer, since it involved carefully planned and extensive experiments which left no doubt about the interconversion of heat and

Fig. 7 James Prescott Joule (1818–1889) whose careful experiments from 1837 to 1847 established beyond question the principle of conservation of energy.

work, and after some doubts convinced the scientific community that energy could not be created or destroyed. A member of a wealthy brewing family, Joule was born in Salford, near Manchester, and received much of his education at home. From 1834 to 1837 he and his brothers took some lessons in mathematics and science from the eminent John Dalton (1766–1844), who had proposed his famous atomic theory in the first decade of the century. None of the Joule brothers had much interest in running the family brewery, and it was sold on the death of their father; they then lived on dividends from the proceeds of the business.

Joule never held any academic or research appointment. All his investigations were carried out in laboratories established at his own expense and installed in the brewery, and afterwards in his various homes. In later life, after experiencing financial losses, he received some subsidies from scientific societies, and during his final years he received a government pension. From 1837, when he was 19, until about 1847, he carried out investigations of various types which led him to conclude that work can be done at the expense of heat, a form of energy, and that the total energy is conserved.

Joule's most important investigations were inspired in 1837 by the work of William Sturgeon (1783–1850), who also worked in Manchester, on electromagnets and electric motors. He was also influenced by an idea that had become prevalent, but which he helped to disprove, namely that there was no limit to the power that could be obtained from a motor operated by an electric battery.

Batteries for the generation of electric current had been developed in the first decades of the nineteenth century, and had led to this idea. Steam engines had to be supplied with fuel obtained from under the ground, and it was realized even in the early nineteenth century that the Earth's supply of fuel was limited. Electric batteries, on the other hand, led to what Professor Donald Cardwell has called an electrical euphoria. One of the enthusiastic proponents of this point of view was Moritz Hermann von Jacobi (1801–1874), a rather remarkable man who had a high reputation in his time, but who is now largely forgotten. He was born in Potsdam, Prussia, but his career was spent further east, first at Dorpat which was then in Russia (it is now called Tartu and is in Estonia), and then in St Petersburg (later Leningrad and now again St Petersburg). In 1834 von Jacobi constructed what was perhaps the first electric motor, and he carried out interesting experiments with it.

In 1835 he published a paper that created something of a sensation at the time. He argued that if certain imperfections of the electric motor, such as friction and what is now called back-emf, could be eliminated, a motor would go on accelerating indefinitely, producing enormous amounts of power. These arguments, although now known to be false, seemed compelling, and many electric motors were built, for a variety of purposes.

At first Joule too was carried away by this electrical euphoria. Beginning in 1837 he carried out careful experiments on the mechanical effect that could be obtained from a motor, and related it to the amounts of metal used up in the battery operating the motor. He was disappointed to find that the consumption of a given amount of zinc in a battery would lead to the production of only about one-fifth of the mechanical work that would be produced by the same weight of coal in a steam engine. Since zinc was much more expensive than coal, this means that an electric motor is far from being a competitor to a steam engine for the primary production of energy. Joule presented this pessimistic conclusion in 1841, in a lecture at the Victoria Museum in Manchester. For him the electrical euphoria was over.

Joule then decided to study the heating effect of an electric current. Using simple equipment, but working with great care, he established that a current passing through a wire of resistance r generates heat in proportion to r, and in proportion to the square i^2 of the current passing; this is his well-known i^2r law. He also concluded that the heat produced is equal to the energy released by the chemical action occurring in the battery. Although this conclusion had later to be modified slightly, it was important in establishing that, contrary to von Jacobi's prediction, energy could not be created from nothing. With regard to the applications of electricity to practical use, Joule wrote that 'electricity is a grand agent for carrying, arranging and converting chemical heat.' This was indeed a shrewd prophesy.

In 1843 Joule carried out an ingenious experiment in which he enclosed the revolving part ('armature') of an electric generator (sometimes later called a 'dynamo') in a vessel containing water, and determined the heat generated when he rotated the armature for a fixed period of time. He also measured the heat produced by the current that was generated. In this way he established the equivalence between the heat produced as a result of rotating the armature and the mechanical work required for the rotation.

In later experiments Joule produced heat in water by stirring it with large paddles. He presented some of his results to the Manchester Literary and Philosophical Society, where the response was quite sympathetic. In 1847 he lectured in the reading room of St Ann's Church in Manchester and the text was reported in detail in a newspaper, the Manchester *Courier*. In this lecture he argued that 'the hypothesis of heat being a substance must fall to the ground'. For good measure, however, he added some metaphysical and religious arguments: 'Believing that the power to destroy belongs to the Creator alone ... I affirm ... that any theory that demands the annihilation of force [i.e. energy] is necessarily erroneous.' The trouble with this type of argument is that another person could apply it to heat rather than total energy.

Since Joule neither was a university graduate nor held a recognized scientific appointment, his work at first met with a chilly reception by the scientific community. Papers he submitted to the Royal Society were rejected. Opinions began to change, however, in 1847 after he attended the meeting in Oxford of the British Association for the Advancement of Science (BAAS, sometimes irreverently called the 'British Ass'), and presented a paper at it. There he had the good fortune to meet William Thomson who soon gave Joule strong support.

William Thomson (1824–1907; Fig. 8), was of Scottish descent, but was born in Belfast, Ireland. He was educated at the University of Glasgow and at Peterhouse, Cambridge. He graduated with high honours in the Mathematical Tripos in 1845, and was at once elected a Fellow of Peterhouse. He then spent a year in Paris working on heat with the distinguished French chemist Victor Regnault (1810–1878). When he first met Joule in 1847 he had been appointed the year before, at the age of 22, professor of natural philosophy at the University of Glasgow; he was to hold that position for over half a century. He is best known today by his title of Lord Kelvin, and to avoid confusion with other Thomsons it seems best for us to call him Kelvin from now on, even though he did not receive that title until late in his life.

Kelvin's scientific work covered a wide range, and he made many contributions of the greatest importance both to science and technology. In his earlier years he was intimately concerned with the new science of thermodynamics, and it was he who first used the word 'thermo-dynamic', in 1849.

Kelvin was impressed by Joule's paper at the BAAS meeting, and had some private discussions with him. By a curious chance, two weeks later Kelvin was on a walking tour in Switzerland and unexpectedly ran into Joule who was carrying a large thermometer; although on his honeymoon, with his bride waiting patiently in a carriage not far away, the enthusiastic Joule was making temperature measurements at the top and bottom of a large waterfall. These meetings, and Joule's papers, finally convinced Kelvin that Joule was right. Heat is a mode of motion,

Fig. 8 William Thomson (1824–1907), who later became Lord Kelvin, and is now usually remembered by that name. He encouraged Joule in his belief that heat is a form of energy, and did important work leading to the second law of thermodynamics. He became a respected figure in science, and was buried in Westminster Abbey beside Newton.

not a substance, and in a steam engine there is an actual conversion of heat into mechanical work. It was probably Kelvin who suggested that from the principle of conservation of energy we can formulate the 'first law of thermodynamics'. This is a more specific formulation of the principle; it states that the change in the total energy of a system is equal to the work done on it plus the heat supplied to it. In using this relationship the same unit must be used for energy, heat, and work. Joule has the signal honour of having the official (Système Internationale d'Unités, or SI) unit of energy, work, and heat named after him. The joule (symbol J) is the energy expended when a force of 1 newton (symbol N) acts through a distance of 1 metre.

Joule had a shy and modest disposition, and his health was always delicate. He married in 1847 but his wife died in 1854 leaving him with two children. He always resided in the neighbourhood of Manchester, and died in Sale, near Manchester, after a long illness.

It is convenient at this point to consider some other aspects of energy, and we should first go back more than a couple of centuries to see what was being done in the field of mechanics. Contributions of great importance to the understanding of the motion of bodies were made by the Italian Galileo Galilei (1564–1642). At

Pisa, and later at Padua, he carried out many experiments on moving bodies and developed mathematical equations to interpret their motion. He allowed a ball to roll on a table with uniform speed until it came to the edge and fell in a curved path to the floor. He realized that the horizontal and vertical motions are independent of one another, and that after the ball left the table its horizontal motion remained unchanged.

Galileo considered also the case of a ball rolling down an inclined plane and then along a horizontal plane. He concluded that the ball tends to continue along the horizontal plane without coming to rest, and this he attributed to its 'inertia'. There is a well-known story that Galileo did some experiments in which he dropped balls from the top of the Leaning Tower of Pisa, but this seems not to have been the case.

Isaac Newton also carried out important experiments on moving bodies, and his ingenious and original mathematical analysis of them led him to formulate three laws of motion. These three laws allowed Newton to make his great contributions to understanding the movement of the Moon and the planets. He invented a new branch of mathematics which he called 'fluxions' and which is a version of differential calculus. Newton's mechanics was developed in detail in his great book *Philosophiae Naturalis Principia Mathematica* which appeared in 1687. Many important topics are included in this book. For example, he described experiments in which bodies were dropped from the cupola of St Paul's Cathedral in London, measurements being made of the times it took them to land. It is curious that these experiments are today not often mentioned, whereas almost everyone 'knows' the apparently untrue story about Galileo and the Tower of Pisa.

Although much more work in mechanics remained to be done, the mechanics described in the *Principia* brought the subject to the end of an important phase. The expression 'Newtonian mechanics' is often used to refer not just to the mechanics developed by Newton but to any used before the introduction of quantum theory; this is more appropriately called classical mechanics.

From our point of view it is interesting that Newton never appreciated the importance of the quantity that we today called energy; he was able to do much mechanics without it. The reason that he did not identify energy is that he did not realize—nor did anyone for several decades—that energy has the special property of being preserved. It cannot be created from nothing, and cannot disappear. Some consideration had been given by earlier investigators into the question of what physical quantity is conserved when collisions occur between spheres. Suppose, for example, that a sphere such as a billiard ball of mass m and speed v collides with another ball. The famous mathematician and philosopher René Descartes (1596–1650) thought about this problem and concluded that the quantity mv (which we now call the momentum) would be conserved; in other words, for two colliding spheres the sum $m_1v_1 + m_2v_2$ is the same before and after the collision. However, his reasoning was partly metaphysical and therefore not convincing. On the other hand the distinguished Dutch physicist and astronomer Christiaan Huygens (1629–1695) carried out many ingenious experiments on colliding spheres, some of which he demonstrated in London at meetings of the Royal Society. These led him to the correct conclusion that the quantity that is preserved in a collision is not mv but mv^2. Similar ideas were expressed by Émilie, Marquise du Châtelet (1706–1749), a woman of remarkable ability who is notorious for having been Voltaire's mistress, and noteworthy for her excellent translation into French of Newton's *Principia*.

This conclusion seems to have become generally accepted by the middle of the eighteenth century, but mv^2 was variously referred to as *vis viva* (literally, vital force), force, impetus, and power. This is confusing to us today, since force, impetus, and power now have precise meanings in science, and must be distinguished from what we now call energy. The word force was once used very loosely to mean any property that could produce some effect; for example, W. R. Grove in his book *Correlation of Physical Forces* (1846) used force in this sense and he even referred to heat as a 'force capable of producing motion'. Today force is defined in science as the rate of change of momentum, momentum being mass multiplied by velocity.

An important modification was made in 1829 by the French physicist and astronomer Gustave Gaspard de Coriolis (1792–1843), who suggested that *vis viva* should be defined not as mv^2 but as $\frac{1}{2}mv^2$. The basis for this definition is that it can be shown by simple mechanics that the work required to give a velocity v to a mass m is $\frac{1}{2}mv^2$. This alternative definition therefore means that work can be regarded as a form of *vis viva*, and that *vis viva* is preserved. For simplicity we will from now on always use the modern word energy to include *vis viva* and the various other forms of energy that are now distinguished.

Today the scientific study of energy is recognized as of great importance, but it is difficult to produce a precise definition of energy in simple language. It is correct to say that energy is either work or anything that can be converted into work. But what is work? For our purposes it is sufficient to say that work is done whenever there is movement of a body against a resisting force. Work is only done when something moves. If there is a heavy carton on the floor and we push against it but fail to move it, we are doing no work on the carton, whatever we may think about all our effort. If we push harder and are successful in moving it we are then doing work, against the resisting force of friction. If in moving the carton we are exerting a constant force and push it through a certain distance, the scientific definition of the work is that it is equal to the force multiplied by the distance.

The reason that the concept of energy was rather elusive, and only clearly recognized in the nineteenth century, is that unlike other physical quantities like pressure, volume, and temperature, it is not directly observable. We always have to infer the energy of an object from some of its characteristics, for example from its temperature or its capacity to do work.

The matter of the different forms of energy is also somewhat elusive and a little confusing, particularly as there are different systems of classification. From the standpoint of classical physics there are just two forms of energy, kinetic and potential. Kinetic energy is energy that a body has by virtue of its motion, and we have seen that for a body of mass m and speed v it is now defined as $\frac{1}{2}mv^2$. Potential energy is energy that a body has by virtue of its position. An example of potential energy is *gravitational energy*, which results from the force of gravity. Thus if a body of mass m is raised from the ground to a height d, we say that its potential energy or gravitational energy has increased by mgd, where g is a factor known as the *acceleration of gravity*. Its value is much the same wherever we are on Earth, but if we go to the Moon it is much less, because there the gravity is much less, because the Moon is so much lighter. The product mg is actually the weight, or force, exerted by the mass when it is subjected to the gravitational attraction of the Earth. This potential energy mgd is equal to the *work* we have to do in raising the body to that height d.

Force and potential energy are also closely linked together: when there is a varying potential energy there is a force. The gravitational force, for example, is due to the fact that the gravitational energy varies with the height from the ground.

Energy can also be classified in other ways. For example, it is sometimes convenient to speak of electrical energy, chemical energy, and nuclear energy, but like gravitational energy these are special types of potential energy. In thermodynamics we speak of the internal energy of a system, and this is a little more complicated since it is a combination of potential energy and the kinetic energy of the atoms contained in the system.

Long before the concept of energy was appreciated, it was realized that there was some property that could not be created from nothing. Over the centuries many devices, some of them highly ingenious, had been invented which were supposed to go on operating for ever, without help from any outside agency. A hypothetical (and non-functional) device of this sort is now referred to as a *perpetual motion machine of the first kind* (we shall meet another kind of non-functional perpetual motion machine in Chapter 4). It was slowly realized, however, that such machines never work. In 1775 the French Académie des Sciences passed a resolution that it would no longer consider any machine claiming to exhibit perpetual motion. A little later, in 1783, the French engineer Lazare Nicolas Marguerite Carnot (1753–1823), in his *Essai sur les machines en géneral*, suggested explicitly that *vis viva* (energy) is conserved and cannot be created. Decades after that, however, it was still maintained by many competent investigators that ways might be found to create unlimited amounts of energy from nothing, as we have already seen earlier in this chapter.

By the middle of the nineteenth century the concept of energy was fairly well understood, and particularly on account of the work of Mayer and Joule heat was recognized as a form of energy, resulting from motion. The theory of heat was often referred to as the 'dynamical theory of heat' or the 'mechanical theory of heat', which is somewhat unsatisfactory, since from one point of view mechanical is just what heat is not. What was implied by these expressions was that heat should be understood as relating to the science of mechanics or dynamics. It is interesting that although we now accept this view we still find it convenient to use language that implies that heat is a substance. We speak of heat 'flowing' from one body to another, and we refer to the 'quantity of heat' in a body. It is difficult to see how we could do differently. In any case, when we speak of the quantity of heat we mean the amount of energy involved.

Having derided Mayer and Joule for so long about the conservation of energy and the nature of heat, the scientific community later held Mayer and Joule in high esteem. The honour accorded to them, however, was tempered with controversy. In 1848 the two men became involved in a bitter priority dispute, carried out mainly through the French Académie des Sciences, with other scientists taking sides regrettably on nationalistic lines; the Germans tended to support Mayer and the British tended to support Joule. An exception was the British physicist John Tyndall, who supported Mayer. The question of priority seems of little importance today and is impossible to resolve since there were no crucial experiments that led inevitably to the conclusion that energy is conserved; the conclusion was seen by both Mayer and Joule to be a probable one in view of the results of experiments relating to the interconversion of heat and work. Mayer's paper of 1842 does give him chronological priority. However, Joule's evidence, based on his numerous careful experiments carried out shortly afterwards, was more thorough and reliable than the less direct evidence adduced by Mayer from the experiments of others.

Convincing expositions of the principle of conservation of energy were put forward in the middle of the century, particularly by William Grove and Hermann von Helmholtz. William Robert Grove (1811–1896; Fig. 9) had an active legal career as well as a scientific one. He went to Brasenose College, Oxford, at a tender age and obtained his BA degree in 1830 at the age of 19. He then practised as a barrister for a period, at the same time carrying out some scientific work. He is particularly remembered today for designing some interesting batteries, including in 1839 a fuel cell that was more than a century ahead of its time, and on which development work is being done to this day.

On suffering some health problems he accepted in 1841 the position of professor of experimental philosophy at the London Institution, which was a short distance from the heart of London, at Finsbury Circus (the London Institution no longer exists, but the building still stands, now occupied by the School of Oriental Studies of the University of London). Scientists find it puzzling that Grove, although he pursued his scientific career vigorously, apparently believed it to be less physically demanding than a legal one. In 1846 he published his book *On the Correlation of Physical Forces* in which he made a clear statement of the principle of conservation of energy (remember that at the time there was confusion about the exact meanings of the words 'force' and 'energy', which today are regarded as quite distinct; when Grove talked about force he meant what we now call energy). He resigned his professorship in 1846 and resumed his legal career, but always retained an interest in science. He became a Queen's Counsel in 1853, and in 1856 was one of the counsel who unsuccessfully defended the notorious William Palmer of Rugeley who was convicted and hanged for the murder by poisoning of

Fig. 9 William Robert Grove (1811–1896), who combined a scientific and a legal career, and was one of the first to state the principle of conservation of energy.

an associate (he also undoubtedly killed several creditors, his wife, an uncle, four of his legitimate children, and several of his many illegitimate children, but was not charged with those murders). Palmer committed his murders by the administration of antimony, strychnine, or prussic acid, and sometimes to make sure by a combination of all three. Grove was presumably chosen as counsel because of his chemical knowledge; his cross-examination of the prosecution witnesses related to the administration of poisons. Grove became a judge in 1871 and was knighted in 1872.

The great German physicist and physiologist Hermann Ludwig Ferdinand von Helmholtz (1821–1894; Fig. 10), who later became a close friend of Kelvin, did outstanding work in both physiology and physics. In 1847 he published a book *Über die Enhaltung der Kraft* in which he dealt with the conservation of energy in a very comprehensive way. (Note again the use of the word *Kraft*, force, to mean energy.) He discussed in detail the dynamical theory of heat, and showed that in inelastic collisions the energy that is apparently lost is converted into heat. This publication probably did more than any other to lead investigators to accept the modern position with regard to energy, heat, and work. It is interesting that when the now famous Cavendish professorship of physics was founded in 1871, the post was first offered to Kelvin, who declined it because he was well satisfied with his professorship at Glasgow. It was then offered to Helmholtz, who also declined it since he had just accepted the chair of physics at Berlin. It was finally offered to and accepted by James Clerk Maxwell, whose distinguished work on energy and other topics we will meet in later chapters.

Fig. 10 Hermann von Helmholtz (1821–1894), the German physiologist and physicist who played a role of great importance in the development and application of thermodynamics. In Germany he was held in great popular esteem, becoming known as the 'Reichchancellor of German physics'. He has the perhaps unique distinction of having been considered for, but refused, professorships at both Oxford (in 1865) and Cambridge (in 1871).

We should note that a qualification to the principle of conservation of energy is required in view of Einstein's special theory of relativity. According to this theory, mass m and energy E are interconvertible, the relationship between the two being $E = mc^2$ where c is the speed of light. When nuclear transformations occur this relationship assumes great importance, but in ordinary chemical reactions the mass changes are too small to detect. We discuss this famous equation and its consequences in Chapter 8.

THREE

Steam engines revisited

It is easy to see the advantages of high-pressure machines over those of lower pressure. This superiority lies essentially in the power of utilizing a greater fall of caloric. The steam produced under a higher pressure is at a higher temperature.

Sadi Carnot, *Réflections sur la puissance motrice du feu*, 1824

It is natural to suppose that the first law of thermodynamics, based on the simple idea of the indestructibility of energy, would have been discovered before the second law, which is concerned with much more subtle aspects of energy transformations. The situation is in reality not quite so straightforward. Instead, the two laws grew up together, and it is only for clarity and convenience that the first two chapters of this book have focused exclusively on the first law. At the same time that the first law was being developed, some significant work was being done which led to formulations of the second law at just about the same time that the first law was being generally accepted. The second law is the subject of the present and subsequent chapters, and we start by considering the remarkable work of a young French military engineer, Sadi Carnot.

Nicolas Leonard Sadi Carnot (1796–1832; Fig. 11), was born in Paris and was a member of a distinguished family. In the last chapter we briefly met his father, Lazare Nicolas Marguerite Carnot (1753–1823), who is important in the history of mechanics as the author of *Essai sur les machines en général* (Essay on machines in general), published in 1783. In this book he discussed the principle of conservation of energy, even though at the time there was little evidence for it. He was also prominent in political and military circles, being called the '*organisateur de la victoire*' because of his administrative activities under the First Republic of France. Sadi Carnot's nephew, Marie François Sadi Carnot (1837–1894), was also a prominent man; he became President of the Third Republic of France in 1887.

Sadi Carnot was educated at the École Polytechnique as a military engineer, and saw active service in 1814. For a few years he held various routine military positions, but being bored by the nature of the work obtained a protracted leave of absence and took up residence in Paris where he undertook study and research in science and engineering. In 1824, when he was 28 years of age, he published a 118-page book, *Réflections sur la puissance motrice du feu et sur les machines propres à développer cette puissance* (Reflections on the motive power of heat...). This book, Carnot's only publication, was of great importance; Lord Kelvin described it as an 'epoch-making gift to science', and another distinguished

Fig. 11 Nicolas Leonard Sadi Carnot (1796–1832), famous for his analysis of the functioning of the steam engine, a study which led to the second law of thermodynamics. (From a painting by Louis Léopold Boilly, done when Carnot was aged 17 and was in the uniform of the École Polytechnique in Paris.)

physicist, Sir Joseph Larmor (1857–1942), described it as 'perhaps the most original in physical science'. It was clearly written but by no means a popular account; it presupposed some knowledge of steam engines, physics, and basic mathematics.

In this book Carnot developed a highly original treatment of heat engines on the basis of his belief in Lavoisier's *calorique* theory. Believing that heat is an imponderable fluid he thought that when heat flows from a higher to a lower temperature and work is done, the heat is actually conserved. Carnot discussed the analogy of a waterfall causing a wheel to turn (Fig. 4 on p. 8); there is no loss of water, the wheel being turned by the force of the water. In the same way he thought that the force of falling heat would cause the piston in an engine to move and perform work. We now know that when work is done as a result of a flow of heat, some of the heat is converted into an equivalent amount of work. We might have thought that since Carnot based his arguments on an incorrect idea his conclusions would be false. Luckily this did not turn out to be the case; his arguments could later be translated into terms based on the idea that heat is not a substance but a form of motion, and his conclusions remain valid.

Carnot was the first to focus attention on the fact that a steam engine cannot function if every part of it is at the same temperature. He considered ideal types of engines which involved a furnace at a higher temperature T_h and a cooler condenser at a temperature T_c. In other words, the engine operates between a higher temperature T_h and a lower one T_c. He imagined a gas to undergo a cycle of changes between the two temperatures, returning to its initial state but performing mechanical work.

One important contribution made by Carnot in his book was that he used for the first time the idea of *thermodynamic reversibility*. (He did not himself use the *word* 'reversibility', which was suggested much later by Peter Guthrie Tait.) This has a rather special meaning. When in thermodynamics we say that a process proceeds reversibly we do not just mean that the process can go in either one direction or the other, which many processes can do. For a process to be thermodynamically reversible it must be reversible *at every stage* of the process. For example, if a substance is being cooled reversibly, the external temperature must at all times be infinitesimally lower than the temperature of the substance and therefore the cooling must be infinitely slow. For any kind of process there cannot be reversibility unless it occurs infinitely slowly. It follows that a thermodynamically reversible process can only occur in our imagination, since if it is infinitely slow it does not go at all. Any process that really occurs must be irreversible in the thermodynamic sense, although if it is slow it may be close to reversible.

Carnot first imagined the gas to be undergoing the cycle of operations completely reversibly; then he imagined it to go round the cycle by some irreversible processes. He was the first to realize that if an engine operates completely reversibly the work it does is the maximum amount that it can do, when it operates between the same temperatures. A consideration of the completely reversible engine is thus useful in giving the *maximum amount* of work that could possibly be done when an engine consumes a given amount of fuel such as coal. Modern treatments of the Carnot cycle usually deal with the *efficiency* of the system, which is the fraction of the heat absorbed at the higher temperature that is converted into work; Carnot, of course, believing in Lavoisier's caloric theory, did not consider that the work was done at the expense of heat that disappeared.

Instead, Carnot considered the maximum 'duty' of an engine, which is the amount of work it does when a given amount of fuel is consumed. He expressed the work done as the mass of water that could be lifted multiplied by how high it is lifted. One important conclusion he reached is that if the engine operates between a higher temperature T_h and a lower temperature T_c, the duty is larger the larger the difference $T_h - T_c$ between the two temperatures, just as with a waterfall (Fig. 4) the greater the fall of the water the more work is done by the water wheel. Carnot also found that for a given drop in temperature, $T_h - T_c$, the work is greater the smaller is T_h. Thus if there is a drop from 1 °C to 0 °C an engine will produce more work than if there is a drop from 100 °C to 99 °C.

Carnot discussed his conclusions with reference to some of the steam engines of his time. He was able to explain why a high-pressure steam engine is more efficient than a low-pressure one. It is simply because the steam generated at a higher pressure is at a higher temperature, so that $T_h - T_c$ is larger.

Another important conclusion reached by Carnot in his book is now referred to as *Carnot's theorem.* He considered two engines, both working between two particular temperatures. One of the engines worked reversibly, and the question he asked was whether the other could be designed in such a way as to produce more work from the same amount of fuel. He answered the question by first postulating that there could be such an engine, and he caused it to drive the first one backwards. He then showed that if this were to occur there would be a net flow of heat from the lower to the higher temperature. This, he pointed out, is contrary to experience; heat always flows from a higher to a lower temperature.

This theorem had great practical implications, which Carnot pointed out. Previously it had been thought that engines could be improved without changing the working temperature, by using different materials; perhaps a steam engine could be improved by changing from water to alcohol or oil, or some other material. Carnot had shown, however, that attention must be directed to the working temperature and not to the materials used.

In 1827 Carnot was required to return to active duty, with the rank of captain, but after less than a year's service he was allowed to return to Paris. He continued his studies on the theory of heat and the design of engines, but made no further publications. Carnot's health was always fragile, and he died at the early age of 36. The exact circumstances of his death are somewhat clouded in mystery. The official version, announced at the time, is that an attack of scarlet fever in June 1832 undermined his constitution, and that in August he fell victim to a cholera epidemic and died within a day. At the time a disease like cholera tended to be regarded as something of a disgrace; nearly thirty years later Queen Victoria's husband Prince Albert died of cholera and that too was hushed up. In Carnot's case there is also more recent evidence that he may have died in a hospital for the mentally disturbed near Paris, a circumstance that would also have been concealed by his family. We do know that after Carnot's death most of his personal papers were destroyed, which was the practice at the time with cholera victims. Those of his papers that did survive suggest that he was abandoning the view that heat is a substance, and was beginning to favour the idea that heat is a form of energy. Had he lived more than his 36 years he would perhaps have made further important advances.

Carnot's book did not at first exert much influence on engineers or scientists. It soon went out of print, and copies were almost impossible to obtain. In 1845 the 21-year-old Kelvin went to Paris to work with Henri Victor Regnault (1810–1878), and tried to find a copy of it. To his surprise and disappointment there was no copy in the library of the École Polytechnique, and no Paris bookseller had heard of it or its author. Kelvin had first learned of Carnot's work from a paper published in 1834 by the French physicist Benoit Clapeyron. Today Carnot's book is easily available, through reprints and translations.

Benoit Pierre Émile Clapeyron (1799–1864) was born in Paris and had known Carnot when they were both students at the École Polytechnique. He became an engineer of some distinction, specializing in the construction of metal bridges and of locomotives. He published little in pure science, but the 1834 paper in which he developed Carnot' ideas caused his name to be remembered by physicists today in two closely related equations, the Clapeyron equation and the Clapeyron–Clausius equation, concerned with the effect of temperature on vapour pressure. But an even more important contribution made in that paper was that Carnot's treatment was restated in the much more precise mathematical language of the calculus.

Carnot had been dead for nearly twenty years before the importance of his work began to be appreciated by the scientific world. One of the first to develop Carnot's ideas was Kelvin, whose work in connection with Joule's experiments on the conservation of energy we met in the last chapter. We saw there that Kelvin, like most other scientists, was at first sceptical about Joule's conclusions but after a few years accepted them enthusiastically. Paradoxically, the reason for his initial reluctance was that he had studied heat conversion more deeply than Joule had; in scientific research there is occasionally an advantage in not knowing too much about a subject. Kelvin knew that if there was an interconversion of heat and work, there was something not entirely straightforward about it. Work could be converted into heat without any apparent complications, as in Rumford's and Joule's experiments, but there were obviously some restrictions on the conversion of heat into work, as had been shown by Carnot. Kelvin grappled with this problem for some time, and was led to a deeper understanding of the restrictions and to what is now known as the second law of thermodynamics.

Kelvin realized, from his study of Carnot's work, that when an engine operates, all of the heat absorbed cannot be converted into mechanical work. Some heat must also simply pass from a higher temperature to a lower temperature, and Kelvin referred to this as the *dissipation of energy*. He saw that it follows that an engine cannot operate at a single temperature. For example, a ship cannot propel itself by abstracting heat from the surrounding cold water; the heat must be obtained from something at a higher temperature, and there must be dissipation of heat, some heat passing from the higher to the lower temperature.

In 1848 Kelvin suggested that Carnot's ideas could be understood more clearly in terms of a specially defined temperature, called the absolute temperature. The temperatures with which most of us are accustomed today are the nearly identical centigrade and Celsius scales, the main exception being the United States which still adheres conservatively to the old Fahrenheit scale, just as it does to the Imperial system of weights and measures. The centigrade scale is based on the freezing and boiling points of water at 1 atmosphere pressure; the freezing point is called zero degrees centigrade (0 °C) and the boiling point 100 hundred degrees centigrade (100 °C). The Celsius scale is in practice almost exactly the same, the modern definition of it being based on two properties that do not involve the pressure. One is the absolute zero, which is a theoretical lowest temperature; we can come close to it but it can never be attained. The other is what scientists call the triple point of water, at which temperature ice, liquid water, and water vapour are in equilibrium. Since the two scales only differ in the second decimal place they are the same for most of our purposes, so that it hardly matters whether we say centigrade or Celsius. In scientific circles, however, it is now considered rather *passé* to say centigrade!

Kelvin made his proposal of a new temperature scale in a paper presented to the Cambridge Philosophical Society entitled 'On an absolute thermometric scale founded on Carnot's theory of the motive power of heat, calculated from Regnault's observations'. This title is somewhat misleading, and has led modern textbook writers to suggest that the scale is based on the efficiencies of Carnot engines (which we discuss in the next chapter). Kelvin changed his ideas while developing this topic, and finally realized that it would be impractical to base the scale on Carnot efficiencies, since these cannot be measured sufficiently accurately. Instead he based them on measurements by the French physicist Victor Regnault on the work done by an engine when a gas expands reversibly. The theory of gas expansion showed that the work is proportional to a temperature defined as the Celsius temperature plus a constant temperature which we now take to be 273.15 degrees.

This temperature is now called the Kelvin temperature or the absolute temperature and, in honour of Lord Kelvin, the symbol used is K. The temperature 25.0°C, for example, is 25.0 + 273.15 = 298.15 K. (By convention we now do not put the degree sign before the K.)

In 1851 Kelvin stated that

> it is impossible ... to derive mechanical effect from any portion of matter by cooling it below the temperature of the coldest of the surrounding objects.

This is one statement of what has come to be called the *second law of thermodynamics*. The year previously the German physicist Clausius had proposed a better and more general definition, which we consider in the next chapter.

FOUR

The second law of thermodynamics

> The second law of thermodynamics holds, I think, the supreme position among the laws of nature. If someone points out to you that your pet theory of the universe is in disagreement with Maxwell's equations—then so much the worse for Maxwell's equations. If it is found to be contradicted by observation—well, these experimentalists do bungle things sometimes. But if your theory is found to be against the second law of thermodynamics I can give you no hope; there is nothing for it but to collapse in deepest humiliation.
>
> Arthur Eddington, *The Nature of the Physical World*, 1927

Ideas similar to those put forward by Kelvin, but expressed in a more precise form, were put forward at about the same time by the German physicist Rudolf Julius Emmanuel Clausius (1822–1888; Fig. 12). Clausius was born in Köslin, then a town in Prussia but now in Poland and called Koszalin. He was educated at the Universities of Berlin and Halle, obtaining his doctorate at the latter university in 1847. In 1855 he became professor of mathematical physics at the Polytechnicum in Zurich, moving to the University of Würzburg in 1867 and the University of Bonn in 1869. He remained at Bonn to the end of his life, serving in his later years as Rector of the University (or, as we would say in most English-speaking countries, its Vice-Chancellor or President).

Clausius was not an experimentalist, and worked mainly on theories of thermodynamics and kinetic theory, but much of his work had significant practical implications. He carried out particularly important research in thermodynamics in the 1850s and 1860s, his main contributions being a detailed analysis of the Carnot cycle and a description of a new physical property which in 1865 he called the *entropy*.

Clausius wrote rather obscurely, and even some highly competent mathematicians and scientists like Kelvin were unable to understand his publications. One person who did manage to understand Clausius's main ideas was Maxwell, who explained them clearly in later editions of his book *Theory of Heat*; even he, however, got things mixed up the first time he explained them, and had to apologise in the next edition. For the time being we will be content with saying that entropy is a property that helps to provide a numerical measure of the extent to which the heat in a system is unavailable for conversion into mechanical work. Later in this chapter, and in later chapters, we will go into the significance of entropy in more detail. We will first look in a general way at Clausius's main publications on the subject.

Fig. 12 Rudolf Julius Emmanuel Clausius (1822–1888), distinguished for his work on the second law of thermodynamics and on the kinetic theory of gases. He introduced the quantity known as the entropy.

Like Kelvin, Clausius only knew about Carnot's work at second hand, gaining his information from the papers of Clapeyron and Kelvin. His first paper on the subject was published in 1850 and had a title that can be translated as 'On the motive power of heat and the heat laws that can be deduced'. In this paper there was an analysis of Carnot's arguments about heat engines, in terms of the amounts of heat absorbed and rejected at the higher and lower temperatures of the engine. This paper is usually considered to be the first to present the second law of thermodynamics, although it was not expressed very explicitly or clearly.

In 1854 Clausius published another paper in which he presented a more detailed analysis of the Carnot engine. For a Carnot engine, operating reversibly and therefore infinitely slowly between two absolute temperatures T_h and T_c, he showed that there is a simple equation for the efficiency of the engine, defined as the ratio of the net work done in the cycle to the heat absorbed at T_h. The equation is

$$\text{Carnot efficiency} = \frac{T_h - T_c}{T_h} = 1 - \frac{T_c}{T_h}.$$

We interpret this equation as follows. The amount of heat absorbed at the higher temperature is proportional to T_h, and the amount of heat rejected at the lower temperature is proportional to T_c. The net heat absorbed is therefore proportional to $T_h - T_c$. In an ideal engine the work done by the engine is equal to the net heat absorbed, so that the ratio of the work done to the heat absorbed at the higher temperature is given by the above expression. In this paper Clausius first introduced the idea of entropy, although he did not yet use the word; as we will see, that was to appear in 1865. He also gave a clearer formulation of the second law of thermodynamics, as follows:

> Heat can never pass from a colder to a warmer body without some change, connected with it, occurring at the same time.

Consider some examples of Carnot efficiencies, remembering that these correspond to the engine working infinitely slowly and that the sole point of working them out is that they tell us the maximum possible efficiency. Suppose first that the lower temperature is 27°C (300 K) and the higher is 100°C (373 K). The efficiency of the reversible engine is then $(373 - 300)/373 = 0.2 = 20$ per cent. In other words, only one-fifth of the heat taken in at the higher temperature 300 K can possibly be converted into work; four-fifths is dissipated, or wasted, by simply passing from the higher to the lower temperature. In practice, since an actual engine cannot be reversible (because then it would operate infinitely slowly) the efficiency will be even less than this.

If instead the higher temperature is 400 K, perhaps because of the use of high-pressure steam, the efficiency is raised to $(400 - 300)/400 = 25$ per cent. Now one-quarter of the heat absorbed at the higher temperature 400 K has been converted into work; the wastage or dissipation is three-quarters, which is less than four-fifths. It will be remembered that Carnot had made the important point that a higher efficiency may be achieved by the use of higher temperatures. Towards the end of the nineteenth century this conclusion was put into practice by Rudolf Diesel (1858–1913), who designed an engine, named after him, in which T_h is much higher than in a steam engine because of the use of oil instead of water. In designing his engine Diesel gave full attention to thermodynamic principles, which he expounded in 1893 in his *Theorie und Konstruktion eines rationallen Warme-Motors* (Theory and Construction of a Rational Heat Engine).

It is useful to calculate these Carnot efficiencies, since by giving us maximum values they give us some idea of what we may expect in practice from an engine. But we must remember that the Carnot efficiency is really the efficiency of an engine that is not working at all. Suppose that we had a steam engine all fired up but not operating, in the sense that it is not doing any work. Since the piston was moving infinitely slowly (i.e. not moving at all) we could proudly boast that the engine was working at top efficiency. One is reminded of a hospital in the TV series *Yes, Minister* that had won an award for efficiency; it had taken in no patients and everything had therefore run perfectly smoothly. Of course, if an engine is working infinitely slowly its power output is zero, since power is defined as the rate of doing work. It is interesting to ask a different question. Instead of asking what is the efficiency when the power output is zero, for that is what the Carnot efficiency is, we can ask: what is the efficiency when the engine is working at maximum power? Engineers have looked into this question, and a simple theoretical treatment gives the result that

$$\text{Maximum power efficiency} = 1 - \left(\frac{T_c}{T_h}\right)^{1/2}.$$

In other words, to calculate this efficiency, we subtract from unity not the ratio T_c/T_h but its square root. Suppose that we consider an engine working between 27 °C (300 K) and 100 °C (373 K), for which we found the Carnot efficiency to be $1 - (300/373) = 1 - 0.8 = 0.2 = 20$ per cent. The maximum power efficiency is $1 - (300/373)^{1/2} = 1 - 0.9 = 0.1 = 10$ per cent. This is less, but at least we are now thinking of an engine that works rather than one that goes infinitely slowly. Of course an actual engine can have a higher efficiency than this, since we may be willing to operate it more slowly, which means at lower power. The efficiency, however, cannot possibly be greater than 20 per cent.

Hypothetical machines in which the second law is allegedly circumvented are known as *perpetual motion machines of the second kind.* Attempts to get around the second law have always ended in failure. One noteworthy attempt was made by the inventor John Ericsson (1803–1889). Born in Sweden, Ericsson spent some time in England before settling in the United States, becoming a citizen in 1848. In most respects he was a highly competent engineer, constructing locomotive engines and a screw propeller which greatly improved navigation. For the US Navy he designed the *Princeton*, built in 1844, the first warship with a screw propeller and with engines below the waterline. Later he designed for the Navy an armoured ship that had a revolving gun turret.

Ericsson based some of his ideas on the Stirling engines, which had been developed in the first half of the nineteenth century by an ingenious Scottish clergyman, the Revd Dr Robert Stirling (1790–1878), and his younger brother James. The important feature of these engines, which are still being operated and developed today, is that they contain a displacer or economizer which functions by minimizing the wastage of heat. Ericsson unfortunately misunderstood the function of the displacer, which he called a regenerator, believing that it allowed the heat rejected at the lower temperature to be utilized over again. This, however, is quite contrary to the second law of thermodynamics. He even thought that he could violate the first law also, for in 1855 he wrote that 'we will show practically that bundles of wire [i.e. the 'regenerator'] are capable of exerting more force than shiploads of coal.'

Ericsson built a vessel, named the *Ericsson*, which was fitted with 'caloric engines', in which this rejected heat was supposed to be 'regenerated' and used again. It was hoped that the *Ericsson* would cross the Atlantic, but we are not surprised today to learn that it was a disaster. It completely failed its trials in 1853 and had to be refitted with steam engines. Even these did not improve the unhappy vessel's luck, as it sank to the bottom of the sea in 1854. Ericsson does not deserve too much blame for his mistakes; when he designed the ship the second law was understood by hardly anyone.

There is no reason to doubt Ericsson's honesty, but the same cannot be said of a number of other individuals who at about the same time made proposals which clearly violated at least one of the two laws of thermodynamics. One of the most notorious of these was the American swindler Robert Keeley, who tried to involve prominent men like the famous financier Cornelius Vanderbilt (1794–1877) in his schemes. He claimed to have made revolutionary discoveries about the forces of nature. One of his projects, by which he wheedled millions of dollars out of credulous investors, involved releasing vast amounts of what he called atomic energy (perhaps the first use of that term) from small amounts of water. He demonstrated the effect by running a motor, which actually operated by means of compressed air entering through hidden pipes. He had the effrontery to try to get the eminent physicist Joseph Henry (1799–1878), then director of the Smithsonian Institution in Washington, to support his nefarious schemes, but Henry was not deceived, and denounced them.

Clausius extended his argument leading to his expression for the efficiency of an engine working between just two temperatures. He also considered processes in which there are a range of temperatures, and this was important since it was this that led him to the concept of entropy. In 1865 he published a paper of particular importance in which he at last introduced the word entropy, and introduced the symbol S for it, a symbol that has survived until the present day. The reason for the symbol S is unknown, but Carnot explained his reason for the choice of the word entropy. He said that he had obtained it from the Greek words η τϱοπη (*en trope*) meaning 'in a transformation'. The Greek word τϱοπη can also be translated as meaning a turn-about, or a change of direction, and Clausius evidently had in mind that entropy is the property concerned with giving a direction to a process. Clausius was not only thinking of etymology when he invented the word, since he wrote: 'I have deliberately chosen the word entropy to be as similar as possible to the word energy: the two quantities to be named by these words are so closely related in physical significance that a certain similarity in their names appears to be appropriate.'

The following may help us to understand the idea of entropy. Often we are particularly interested in some kind of a device, and would like to relate the heat evolved to the work that the device can perform. An electric battery provides us with a good example. We have seen in Chapter 2 that Joule was interested in the interconversion of heat and work, and that he made many measurements on electric batteries. He showed that the work that was performed by a battery (e.g. in driving an electric motor) was done as a result of the chemical reaction that occurred in the cell. He concluded, reasonably but unfortunately not quite correctly, that all of the heat released by the chemical reaction would be converted by a motor into mechanical work. The truth turns out to be a little different. As in a steam engine, some of the heat may be unavailable for conversion into work. This is the case if the chemical system undergoes a decrease of entropy as the reaction in the battery occurs. If this is so, the thermodynamic result is that the work done by the chemical system in the battery is the heat evolved *minus* the product of the entropy decrease and the absolute temperature. In other words, there is some unavailable energy, which Clausius showed to be given by unavailable energy = entropy decrease in the system × absolute temperature.

(If instead there is an entropy *increase* in the chemical system, which is possible, the work done by the battery will be greater than the heat that would be evolved—the unavailable energy is negative. This sounds as if we are getting energy for nothing, but we are not; the point is that heat transferred from the environment ensures a true energy balance.)

It is easy to understand why Joule had earlier reached the wrong conclusion, since his experiments were done before the idea of entropy had been suggested. It is interesting that Kelvin made the same mistake even though he based his conclusion on what seemed to be excellent experimental evidence. Kelvin considered a Daniell cell, which was a popular electrical cell (commonly called a battery) used at the time. From the heat evolved in the chemical reaction he calculated what the voltage of the cell should be. His result agreed so well with the experimental value that there seemed no doubt that heat in an electric cell was simply converted into work, with no complications. His error resulted from an unlucky chance. For the particular reaction occurring in that cell, the entropy change is practically zero, so that the unavailable energy is almost zero. This example is instructive in showing us how easy it is to go wrong in drawing a scientific conclusion, even using excellent experimental evidence and sound logic. A certain amount of bad luck may also be involved in choosing an appropriate system on which to experiment.

Clausius concluded that when any spontaneous process occurs—such as a building falling down or an explosion occurring—there will be an increase in the entropy of the universe as a whole. The increase need not be in the system itself; there can be a loss of entropy in the system, but the process will only occur if there is a greater gain in the surroundings, resulting from the emission of heat. In his 1865 paper Clausius expressed the two laws of thermodynamics in a compact form, as follows:

The energy of the universe remains constant.
The entropy of the universe tends towards a maximum.

These two laws are sometimes expressed in everyday language as:

You can't get anything for nothing.
You can't even break even.

Entropy is so subtle a property that many scientists were unable to understand it when Clausius first suggested it. Kelvin, for example, never appreciated entropy, and maintained that the second law can be more easily understood in terms of the dissipation of heat, which is easily visualized. Kelvin's philosophy of science was that everything must be explained in terms of a mechanical model, and entropy cannot be explained in this way. Properties like volume, pressure, and temperature can be measured with simple instruments and can be appreciated even by people who do not know much about science. Entropy, on the other hand, is elusive; no instrument can directly measure an entropy change, which has to be calculated from data involving heat and temperature changes.

Kelvin was wrong to dismiss the idea of entropy, however. The dissipation of heat is by no means as satisfactory as entropy in leading to an understanding of why processes occur. Analysis of the situation shows there is not an exact correlation between the dissipation of heat and the change in entropy, which is the property that does give a precise understanding of the tendency of a process to occur. Only in some simple and special cases is there an exact correlation between the two factors. The mixing of two gases at the same temperature and pressure provides a good example. There is no heat change at all, so that Kelvin's argument about the dissipation of heat cannot provide any help. Clausius's procedure, however, provides a way of calculating the entropy change for the mixing process, which is positive. It therefore explains neatly why the mixing of two gases or liquids occurs, while the unmixing of a mixture is not something that we ever observe.

We saw at the end of Chapter 2 that there were some bitter controversies about the priority of discovering the first law of thermodynamics, some giving credit to Mayer, others to Joule (with poor Rumford generally left out). There were also fundamental and acrimonious disagreements about the second law of thermodynamics, with harsh words written and spoken on both sides. The unpleasantness was precipitated by Peter Guthrie Tait (1831–1901; Fig. 13), whose career was a colourful one. He was born in Dalkeith, Scotland, in the same year as James Clerk Maxwell, whom we will get to know well in later chapters. Tait and Maxwell attended the same school, the Edinburgh Academy, and they both went first to the University of Edinburgh and then to Cambridge, from which both graduated with the highest honours. In 1854 Tait became professor of mathematics at Queen's College, Belfast, and in 1860 he was appointed professor of natural philosophy (which we now call physics) at the University of Edinburgh, holding that position until shortly before his death. Maxwell, incidentally, was also a candidate for that professorship; he was a better scientist than Tait, but not a good lecturer, in contrast to Tait who was excellent.

Tait was over 6 feet (1.83 metres) tall, strongly built, looked like a prizefighter, and had an overpowering personality. He worked in a wide variety of fields, including thermodynamics. For a time he exerted a considerable influence in scientific circles, especially through his textbooks of physics, particularly a textbook which he wrote in collaboration with Kelvin, with whom he was always on friendly terms. This was Volume 1 of *Treatise on Natural Philosophy*, which first appeared in 1867; a promised Volume 2 failed to appear. This book was usually referred to as 'Thomson and Tait' or as 'T and T', and since the Archbishops of York and Canterbury at the time happened to be named Thomson and Tait, Maxwell referred to his two scientist friends as the 'Archiepiscopal Pair'. In 1896 Tait published two memorable papers on the ballistics of a golf ball. Besides giving a very mathematical account of the topic, the papers contained much valuable advice to golfers, and Tait later lamented that golfers were nearly always so set in their ways that they would not change their methods even if it meant lowering their handicaps. In particular, Tait said, golfers were horrified to be told that they

Fig. 13 Peter Guthrie Tait (1831–1901), for many years professor of natural philosophy at the University of Edinburgh, and co-author with William Thomson of Volume 1 of *Treatise on Natural Philosophy*, 1867. (Portrait from C. G. Knott, *Life and Scientific Work of Peter Guthrie Tait*, Cambridge University Press, 1911.)

would do better to put a spin on the ball by slicing it to some extent; to them a golf slice was a disgraceful thing. Eventually Tait's ideas were accepted, and the dimpling of golf balls to improve their ballistics resulted from his suggestions. He himself was a superb golfer, and one of his four sons followed his father's advice and became a well-known champion golfer; he unfortunately lost his life in the Boer War.

During his lifetime Tait was involved in a number of scientific controversies, often rather bitter ones. His controversy with Clausius about some of the fundamental principles of thermodynamics, and about priorities, was a long and stormy one. It started in the middle 1860s and raged on in the pages of the *Philosophical Magazine* and of Poggendorff's *Annalen der Physik und Chemie.* It continued in various editions of their books, specifically in Clausius's *Abhandlungen über die mechanische Wärmetheorie* (Treatise on thermodynamics), which appeared from 1865 to 1867, and in Tait's *Sketch of the History of Thermodynamics*, the first edition of which appeared in 1868 and a second in 1877. On the whole Clausius was fair, but Tait revealed himself as the worst kind of chauvinist. It appears that in his opinion the best work was inevitably done by a Scot (e.g. Kelvin and Maxwell), the work of a man from northern England (e.g. Joule) might also be competent, but little good could come from a Frenchman (e.g. Carnot and Clapeyron) and certainly not from a German (e.g. Mayer and Clausius). As a result Tait gave the credit for the first law to Joule rather than Mayer. He gave Kelvin the credit for the second law of thermodynamics, and discounted Clausius's idea of entropy change, which he said was the same as Kelvin's dissipation of heat. Unfortunately for him his exposition of this was quickly demolished by Clausius.

Worse still, in his own writings Tait even reversed the meaning of entropy. Instead of regarding entropy, as Clausius did, as related to the unavailability of energy, he perversely defined entropy as the availability of energy. His version of the second law was therefore that entropy should decrease in a spontaneous process. In so doing he even confused Maxwell, who in earlier editions of his *Theory of Heat* defined entropy in Tait's way but discussed it in Clausius's way. For spreading such confusion Maxwell apologised in the preface to the fourth edition (1875) and in a mild and amusing note to Tait chided him for getting him into such a mess.

Today these priority disagreements are mercifully all but forgotten except for their entertainment value. As to the priority for the second law, I think that the last word was written by the great American thermodynamicist Josiah Willard Gibbs (1839–1903). Today Gibbs is considered to be perhaps the greatest genius of thermodynamics. However, largely because of his retiring nature and his inability to communicate clearly, he gained little recognition during his lifetime. It has been said that for a time Maxwell was the only one who understood Gibbs's thermodynamics, which he did mainly by working it out for himself. Some prominent scientists found little use for the work; Kelvin, for example, in a letter to Lord Rayleigh wrote: 'I find no light for either chemistry or thermodynamics in Willard Gibbs.' For many years, from 1871 until his death, Gibbs was a professor at Yale University, and the story is told that on one occasion the President of Yale told Maxwell that Yale was looking for a professor of theoretical physics since it did not have one; Maxwell mentioned Gibbs, but the President had not heard of him.

After Clausius died in 1888, Gibbs prepared an obituary notice for him which was published in the Boston-based *Proceedings of the American Academy of Arts and Sciences.* His article is remarkable for including not only an appropriate tribute to Clausius, but an impartial statement of the origins of thermodynamics and an assessment of the contributions of the various founders. It contained the clear statement that the subject of thermodynamics had begun in the year 1850 with the publication of Clausius's first paper on the subject of the second law. He also commented, in an elegantly inoffensive way, that a statement of the second law by Kelvin in 1851 was clearly based on Clausius's ideas in his 1850 paper.

In any case, of course, Kelvin's interpretation of the second law was less satisfactory than that of Clausius, in that he never recognized the significance of the entropy and the important role it plays.

People often refer to a third law of thermodynamics, by which they mean a theorem proposed in 1906 by the distinguished physical chemist Walther Nernst (1864–1941) and known as the Nernst heat theorem. This theorem relates to entropy changes when processes occur close to the absolute zero. It does not have the same broad status as the first and second laws of thermodynamics, and even Nernst himself did not like his theorem to be called the third law. The law is important to those who carry on research at very low temperatures, but need not concern us in the present book.

FIVE

Maxwell's demon

> The 2nd law of thermodynamics has the same degree of truth as the statement that if you throw a tumblerful of water into the sea, you cannot get the same tumblerful out again.
>
> James Clerk Maxwell, letter to J. W. Strutt (later Lord Rayleigh), December 1870

In the last chapter I tried to give the reader some idea of the significance of entropy without using any of the rather subtle mathematics that is required for a complete understanding of it. There is one important aspect of entropy, however, that we have not yet discussed. Just why does the entropy of the universe constantly increase, as the second law of thermodynamics says it must? This turned out to be a rather difficult question to answer. If this were a mystery novel I should go through all the debates about the matter, and only reveal the answer at the end; any other procedure would be considered unacceptable according to the rules of detective fiction. But this is not a mystery book—or is not intended to be—and I think that the best thing is to reveal the answer to the dilemma right away. Later, in this and the next chapter, I will tell something of the debate that took place between some very eminent scientists of the nineteenth century. This might be thought to be a waste of time, since one side of the argument was clearly mistaken. However, I think that the presentation of the arguments in outline does help us greatly to understand some important aspects of what some scientists consider to be the most fundamental law of nature.

The two main answers to the question of why the entropy of the universe always increases as time goes on may be stated briefly as follows. The first, the correct one, is that it is all a matter of probability. There is a natural tendency for systems to pass from orderly states to states of greater disorder just because the disordered states are more probable. The analogy of a deck of cards is often used to explain entropy. A deck can be arranged in a particular way, such as the way the cards are arranged by the manufacturer in an unopened pack. Alternatively, it can be shuffled. What we define as a shuffled or disordered deck is more probable than the ordered one. This is because there is an enormously large number of arrangements that we call disordered, while only a smaller number, perhaps only one, qualifies as ordered. Shuffling an ordered deck will almost certainly produce a disordered deck, and it is highly unlikely that a shuffled deck will become ordered if it is further shuffled. According to this view the second law, unlike the first one, is not an absolute law. It is possible for a system to go from a disordered state to a more ordered one, just as a shuffled deck of cards may, on further shuffling, become an ordered one. Such things are not impossible but unlikely, and we will see how extremely unlikely they are.

The second answer to the question, the one that proved to be wrong, is that nothing more is required than the laws of mechanics. The view taken was that from these laws one could, purely mathematically, reach the conclusion that a process such as the mixing of two gases would inevitably occur in the direction of mixing, and could never move in the opposite direction. On this view a violation of the second law would be absolutely impossible, not just highly unlikely.

The first answer, the correct one, may be expanded as follows. We are all familiar with processes that occur naturally as a result of this tendency for an ordered state to become a disordered one. When a lump of sugar is dissolved in coffee we know that the molecules of the sugar spread themselves throughout the liquid; however long we wait, we do not find the cube reforming itself—although if we could wait a very long time (longer than the age of the universe) it would do so—but at once dissolve again. We know that if a bottle of perfume is left open, the perfume will spread around the room, and we do not expect that in our lifetimes the molecules of perfume will go back again into the bottle. Some of us have seen a demonstration experiment in which oxygen and hydrogen gases are brought together, and a flame is put to the mixture; the gases explode with the formation of water (H_2O). The opposite process does not occur, however. However long we wait, a glass of water will not suddenly decompose into hydrogen and oxygen. The reason is that there is a great increase of entropy, in the gases and the environment, when the gases are exploded together, largely because heat is given off and is dissipated into the surroundings. In principle heat from the surroundings could assemble in a glass of water and decompose it into hydrogen and oxygen, but the probability of this happening is extremely remote.

We should note that time enters into this argument. Disorder increases as time passes. The British philosopher and astronomer Sir Arthur Eddington (1882–1944) looked at the matter from this point of view, and referred to entropy as the 'arrow of time'. Time cannot go backwards but only forwards, and this is because at a later time the state of the universe has a greater probability than at an earlier time.

It is useful to consider the mixing of gases in more detail, and to do so we need to know something of what is called the kinetic theory of gases. This theory is concerned with understanding the properties of gases in terms of the behaviour of the individual molecules. Two simple relationships apply to gases to a good approximation. One is that at a given temperature the product of the pressure and the volume is a constant, so that if for example we double the pressure on a gas its volume is halved. This is known as Boyle's law. The other is that at a given pressure the volume is proportional to the absolute (Kelvin) temperature; this is Gay-Lussac's law, often known as Charles's law. The first serious attempt to explain these and other properties of gases was made in 1738 by the Swiss mathematician Daniel Bernoulli (1700–1782). His idea, illustrated in Fig. 14, was

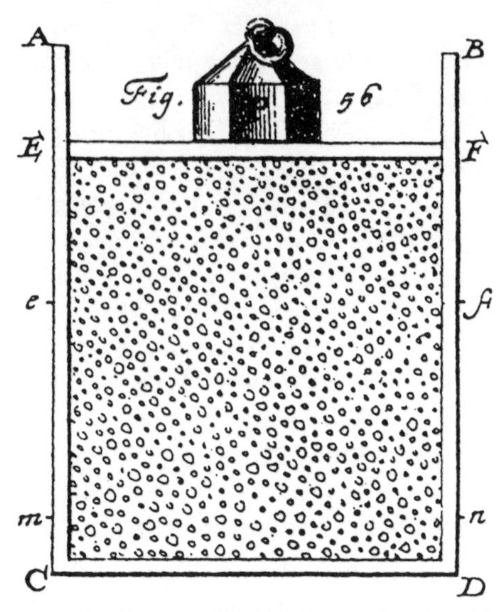

Fig. 14 Daniel Bernoulli's diagram (1738) illustrating his ideas about the kinetic theory of gases. The pressure of a gas is due to the bombardment of the molecules on the surface of the container.

that the pressure of a gas is due to the bombardment of the molecules on the surface of the vessel, This idea, however, was ahead of its time, and although undoubtedly correct was not taken seriously for many years.

It was not indeed until over a hundred years later that much progress was made in developing Bernoulli's ideas. There were, however, two contributions that would have exerted much influence if their efforts had been treated more sympathetically by the scientific establishment of the time. The first of these was made by John Herapath (1790–1868), who was born in Bristol and received little formal education. He did gain some competence in science by working in the business of his father, who was a maltster, and by his extensive reading. While still in his teens he was recognized locally as something of a mathematical prodigy, and he later became a teacher of mathematics. In 1820 he submitted to the Royal Society, for publication in the *Philosophical Transactions*, a manuscript entitled 'A mathematical enquiry into the causes, laws and principal phenomena of heat, gases, gravitation, etc.'. It included among other things a mathematical treatment of the movement of molecules in gases and was based on the same ideas that Bernoulli had proposed, although it appears that he was not aware of Bernoulli's work on the subject. The Royal Society rejected the paper, largely as a result of criticisms made by the famous chemist Sir Humphry Davy who was then exerting a strong influence on the progress of science and was in that year elected President of the Royal Society. Although Davy was himself sympathetic to the idea that heat is a form of motion, his opinion was that Herapath's paper was too speculative, and that it also contained some mathematical and scientific errors. With a more sympathetic response Herapath might easily have removed these errors, but instead he withdrew it from the Royal Society and sent it to another journal, the *Annals of Philosophy*, where it appeared in 1821.

He made a number of subsequent attempts to publicize his ideas; he wrote letters to *The Times*, and published articles on his theory, referred to as ***kinetic theory***, in the *Railway Magazine*. This seems a singularly inappropriate medium for such publications, and one wonders what the typical readers of that magazine made of the mathematical treatments found in them. The reason for Herapath's eccentric choice was that he was a great railway enthusiast, and he later edited the *Railway Magazine* himself. In the field of railway engineering he was able to exert a stronger influence than he had done in science, and sometimes applied his kinetic theory to railway problems. In 1836, for example, he published a treatment of the wind resistance encountered by a fast locomotive in terms of molecular speeds.

One of Herapath's most important achievements was to use his kinetic theory to calculate the speeds of molecules in a gas, and to calculate the velocity of sound in air. He presented this work to the BAAS at its meeting in 1832, but this contribution was soon forgotten. Joule made similar calculations some years later, and is usually given the credit for being the first to do so.

In 1847 Herapath published a book, *Mathematical Physics*, which exerted some influence on a few scientists, including Joule. On the whole, however, Herapath's work was neglected, for various reasons. One was that his basic idea, the same as that of Bernoulli, that the movement of particles is related to heat and pressure, was suspect at the time; there was still general belief that heat is a substance rather than a manifestation of the movement of particles. Also, his detailed mathematical treatment had some faults which critics seized upon. His treatment therefore required a good deal of modification to become entirely satisfactory, but the ideas were essentially correct and important, and with a little help from others he might have been able to produce a correct formulation of the kinetic theory. As it turned out, Herapath was regarded by scientists as a rather eccentric amateur who would do better by confining his attention to railways rather than intruding into science. Later scientists were kinder about his work. Maxwell, for example, while admitting that some of the details of Herapath's work were faulty, wrote that 'his speculations are always ingenious, and often throw much light on the questions treated. In particular, the theory of the temperature and pressure of gases and the theory of diffusion are clearly pointed out.'

The second important contribution to kinetic theory was also rejected in its time by the scientific community. It was made by John James Waterston (1811–1883), who was born in Edinburgh and unlike Herapath received a fairly good education. While apprenticed as a civil engineer he attended lectures at the University of Edinburgh. In the 1830s he practised as a railway engineer and then moved to India where he was in the service of the East India Company and taught naval cadets. It has been suggested that his success with kinetic theory, which involves collisions between molecules, was partly due to the fact that his leisurely life in India allowed him to become a proficient player of billiards which involves collisions between billiard balls. In 1845 while still in India he submitted to the *Philosophical Transactions* a paper which gave an essentially correct and comprehensive kinetic theory, better than that of Herapath. It included for the first time a statement of the equipartition of energy among molecules of different masses, it treated the speeds of the molecules, and it gave a satisfactory treatment of specific heats.

The paper was rejected for publication by the Royal Society, but an abstract of it appeared in the *Proceedings of the Royal Society*. The Secretary of the Royal Society at the time was the physiologist Peter Mark Roget (1779–1869), now best remembered for his *Thesaurus of English Words and Phrases* (1852), and he sent Waterston's paper to two referees. One was the Rev. Baden-Powell (1796–1869), the Savilian Professor of Geometry at Oxford, who had done some experimental work in the field of optics; he was the father of the founder of the scouting movement. The other was Sir John Lubbock, Bart. (1803–1865), a banker and amateur scientist who had done some good work in astronomy and the theory of the tides. Neither referee knew much about the subject matter of Waterston's paper, and both rejected it outright. Baden Powell's objection was that the basic assumption that pressure is due to molecular bombardments is 'very difficult to admit, and by no means a satisfactory basis for a mathematical theory.' Lubbock said that 'the paper is nothing but nonsense, unfit even for reading before the Society.' The paper was, however, 'read'—in the formal sense that its title was read out at a Royal Society meeting—and Roget prepared the abstract that was later printed in the *Proceedings*.

The policy of the Royal Society was that a paper that had been read before it became its property and could not be returned to the author. In Herapath's case he had withdrawn his paper before it was read, and therefore it was returned to him and he published it elsewhere. Waterston, however, perhaps because he was in India at the time and out of touch with the situation, allowed his paper to be read and therefore could not have it returned. In those days there were no photocopiers, and a copy could only be made by hand. Waterston had unwisely made no copy, and the paper did not see the light of day for over 40 years.

Waterston saved money during his stay in India, and in 1857 retired at the age of 46, returning to Scotland and devoting himself to further research. One of his interests was the age of the Sun and the Earth. He was ahead of his time in appreciating that the Earth must be considerably older than it was generally assumed to be. He suggested that the Sun could have been kept at a high temperature for a long period by the fall of matter into it, heating being produced from gravitational energy; this is consistent with modern ideas as we shall see in Chapter 9. Like Mayer, Waterston became embittered and depressed by the neglect of his ideas by the scientific community, and he tended to avoid the society of scientists. On 18 June 1883 he walked out of his house near Edinburgh and vanished into thin air, being presumed to have done away with himself.

It was in 1892, nine years after Waterston's death, that his important paper of 1845 was finally resurrected. In that year John William Strutt, 3rd Baron Rayleigh

(1842–1919), discovered it in the Royal Society archives and had it printed in the *Philosophical Transactions*, with an introductory note by himself in which he expressed regret that this valuable paper had not been printed earlier. By the time the paper appeared the work had all been done again, particularly by Maxwell and Clausius, so that the paper was then only of historical interest. The comment has sometimes been made that it seems surprising that when others developed the same theory those who knew of Waterston's work did not call attention to it. However, those who did know of his work, Roget, Baden Powell, and Lubbock, probably did not follow the developments in kinetic theory, their own scientific interests being elsewhere.

Another contribution to kinetic theory was made in 1848, when Joule presented a paper in which he calculated the velocity of a hydrogen molecule. This had already been done by Herapath, and Joule's paper was not of much significance. More important work on the theory was done by Clausius, who in 1857 and 1858 derived a fundamental relationship between the pressure–volume product for a gas, and the number of molecules, their mass, and their average speed. In 1858 Clausius made a contribution of importance by recognizing a quantity known as the *mean free path* of the molecules in a gas. This work resulted from an objection that had been raised to the fact that he had deduced from kinetic theory that at ordinary temperatures the molecules would be moving at hundreds of metres a second. That brought a quick objection from the Dutch meteorologist Christoph Hendrick Diedrich Buys Ballot (1817–1890). He pointed out that if he were sitting at the end of a long table and someone brought in dinner at the other end, it would be a few seconds before he could smell what he was about to eat. But surely, if the molecules were moving so fast, he would smell the dinner in an instant.

Clausius realized that this objection would only apply if molecules had no size, in which case a molecule could travel from one end of a room to another without being impeded. In fact molecules have a size and there will be collisions between them so that a molecule cannot travel far in a straight line. In travelling from one place to another a gas molecule will therefore undergo many collisions with other molecules and will make slower progress. Clausius's mean free path is defined as the average distance that a molecule travels between two successive collisions, and he worked out the theory that applies to it.

At the same time Maxwell also presented treatments of gases which had particularly important implications for the understanding of the second law, and we should now consider his career and work in a little detail. James Clerk Maxwell (1831–1879; Fig. 15) was born in Edinburgh to a family of comfortable means. His father John Maxwell was Laird of Glenlair, a position that James inherited on his father's death. He also inherited a strong streak of eccentricity. His biographers reported that his grandfather on one occasion escaped from drowning in an Indian river by floating to shore on his bagpipes, whereupon he played them to entertain his companions and to frighten away any tigers. His father too was an individualist, wearing peculiar shoes and clothes made to his own design.

Fig. 15 A photograph of James Clerk Maxwell taken during his tenure of his Cavendish professorship at Cambridge.

Throughout his life James displayed a humorous eccentricity of which he was well aware and often enjoyed displaying. He was always late to bed and a late riser, and on arriving at Cambridge and being told that he would be expected to attend early morning prayers in his college chapel, his reply was 'Ay, I think I can stay up 'til then.' Throughout his life he had a boyish sense of humour, and enjoyed playing harmless and humorous pranks on his many friends.

From an early age James showed great curiosity about the world around him. His favourite comment was 'What's the go o' that?', and if the answer did not satisfy him he would ask 'But what's the particular go of it?' Much of his boyhood and some periods of his adult life were spent at the manor house at Glenlair, which is in the south-west of Scotland. As a boy in that beautiful and rustic neighbourhood he acquired such a broad Scottish brogue that he was often understood only with difficulty even by fellow Scots, particularly those from Edinburgh.

At the age of 10 he was sent to the Edinburgh Academy, where on account of his accent and his homespun ways he was somewhat ridiculed by his fellow schoolboys, who inappropriately nicknamed him 'Daftie'. Maxwell seems not to have minded too much.

In 1847, at the age of 16, he became a student at the University of Edinburgh. He took his Edinburgh degree in 1850 and then became an undergraduate at Cambridge. He was first attached to Peterhouse, but decided that the chances of a fellowship would be greater at Trinity College. After a term at Peterhouse he therefore transferred to Trinity, where he came under the influence of William

Hopkins (1793–1866), an able scientist and excellent teacher. Since he was married Hopkins could not be a college fellow (according the rules at the time), but he became a private tutor, and was remarkably effective in helping students to get good degrees, being called the 'Wrangler-maker'; this refers to the fact that the Cambridge students who got the best (first-class) degrees were called Wranglers. Kelvin and P. G. Tait, and many others, had profited from Hopkins's tutoring. In 1854 Maxwell graduated as Second Wrangler, the Senior Wrangler of the year being his friend Tait.

While still at Cambridge he began an important investigation on the nature of the rings of the planet Saturn. Astronomers had observed three concentric rings about Saturn, all in the same plane. At least some regions of the rings were known to be quite thin, since the planet behind can be plainly seen. Maxwell carried out a careful theoretical treatment, and concluded that the rings could not be solid or liquid, since the mechanical forces acting upon rings of such immense size would break them up. He suggested that instead the rings must be composed of a vast number of individual solid particles rotating in separate concentric orbits at different speeds. His final article on the subject, published in 1859 after he had moved to Aberdeen, is a meticulous and lucid analysis of the problem.

Later studies, including observations from the Voyager spacecraft in the latter years of the twentieth century, confirm Maxwell's conclusions. The fact that the rings are composed of particles is supported by observations of stars seen right through portions of the rings. Spectroscopic studies have shown that the particles are composed of impure ice, or at least are ice covered. Radar observations have confirmed the range of speeds predicted by Maxwell. It appears that the particles have diameters from a few centimetres to about a hundred metres. This is one of the most remarkable examples of a fact being deduced satisfactorily from theory.

In 1856 Maxwell became professor at Marischal (pronounced 'Marshal') College, Aberdeen, but he did not enjoy lecturing and was not much of a success at it. His Scottish brogue was strong, his diction was poor, and he had something of a stutter when he was nervous; as a result he could not hold the attention of his students. He also had an unfortunate tendency, throughout his career, to make mathematical mistakes, being particularly careless (one biographer has said 'cavalier') about the use of plus and minus signs. He was even apt to express his ideas in a confused way. In a man who wrote with such great clarity this is most surprising; even though he was conscientious in preparing his lectures he always became muddled when he started to speak.

In 1858 Maxwell married Katherine Mary Dewar, the daughter of the Principal of Marischal College; she was seven years older than himself. There is no doubt that their married life was one of great mutual devotion. Katherine, however, seems to have been disliked by her husband's relations and friends, of whom there were many, as he was very popular. She seems to have been dour and humourless, and none of them ever had anything nice to say about her. Some were very critical, referring to her as a 'difficult woman'. Katherine Maxwell did not much appreciate her husband's scientific career, preferring to perform the social duties of the wife of the Laird of Glenlair, which they visited as often as possible. She was, on the other hand, of much help to her husband in some of his experiments.

When in 1860 there was some reorganization at the college there was no longer room for Maxwell. One would have thought that being the son-in-law of the Principal of the College would have helped him to retain his position, but it seems possible that it may have worked to his disadvantage. Katherine Maxwell would have much preferred her husband to have relinquished his academic work and taken up permanent residence as Laird of Glenlair. If this is so her scheme was frustrated, as he was determined to obtain another academic post. He applied for the professorship of natural philosophy (physics) at the University of Edinburgh, but the appointment went to his friend Tait who, unlike Maxwell, was a superb lecturer. Maxwell was instead successful in securing a professorship at King's College, London, where he remained until 1865. In that year he retired for some years to his family estate in south-west Scotland. In 1871 he somewhat reluctantly agreed to become the first Cavendish professor at Cambridge, the post having previously been refused by Kelvin and Helmholtz. He held the post for only eight years, dying of abdominal cancer at the age of 48, the same age as his mother when she too died of the same ailment.

Maxwell is particularly noted for his treatment of the distribution of molecular speeds in a gas, and for his theory of electromagnetic radiation. He also made important contributions in other fields. He was a skilful experimentalist, and designed some of the instruments used in the Cavendish Laboratory. He and his wife made the first reliable measurements of viscosities of gases, using apparatus of his design.

Maxwell's work on the kinetic theory of gases was begun while he was at Marischal College, and derived from his work on Saturn's rings. Early in 1859 he saw a translation of one of Clausius's papers, and he set out to develop the theory. At the 1859 meeting in Aberdeen of the BAAS he presented a theory of the viscosity of gases on the basis of his kinetic theory. Viscosity is a property that is related to the difficulty with which a substance flows. When we pour oil it flows more slowly than water, and we say that oil has a higher viscosity than water. Gases have much lower viscosities than liquids, and flow much more freely. When Maxwell got interested in their viscosities, he found that hardly any data had been obtained. His mathematical treatment of the properties of gases in terms of the movement of their molecules led him to conclude that their viscosities should be independent of pressure, and that they should increase approximately with the square root of the absolute temperature. He thought it rather peculiar that the viscosity should not depend at all on the pressure, and that a gas should become more viscous as the temperature is raised. With liquids the opposite is true; warming a liquid causes it to flow more easily, which means that it has a lower viscosity. Maxwell decided that experimental results were needed on gases, and these he obtained after he took up his professorship in London.

In the attic of his house in Kensington, and with the help of his wife Katherine, Maxwell made many experimental measurements of gas viscosities, in order to confirm the conclusions he had drawn about the effects of pressure and temperature. Temperatures between 51°F (10.6°C) and 74°F (23.3°C) were brought about by simply changing the temperature of the attic room; this was arranged by Katherine, who organized the appropriate stoking of the fire and opening and closing of windows. Some work was also done at 185°F (85°C), and this temperature was achieved by a suitably directed current of steam. The results of this investigation, which confirmed Maxwell's predictions from kinetic theory, were published in 1866.

At the 1859 BAAS meeting in Aberdeen Maxwell also announced a theory, for which he has become famous, of the *distribution of molecular speeds*. It had already been recognized that in a gas, as well as in a liquid, some molecules will be moving rapidly and others more slowly. Clausius was well aware of this, and in his papers he had explained, in terms of the variation of molecular speeds, why liquids become cooler when evaporation occurs. The more rapidly moving molecules are more likely to leave the surface of a liquid, so that the molecules left behind after evaporation have on the average less energy; the temperature of the liquid remaining is therefore lower. It is rather curious that although Clausius understood this he never brought the distribution of speeds into his kinetic theory, always using average speeds, which is not entirely satisfactory.

In 1860 Maxwell published a paper, based on his presentation at the Aberdeen meeting, in which he derived an equation for this distribution of molecular speeds in a gas (Fig. 16). He realized that it would be hopelessly complicated to try to deal with the individual motions of a large number of molecules, but that probability theory provides a shortcut to a solution. Probability theory deals, for example, with the problem of the number of ways in which a certain number of balls can be distributed among a certain number of boxes. With characteristic insight, Maxwell realized that molecular distributions could be dealt with in the same way. He considered how molecular speeds would distribute themselves amongst a large number of molecules.

Maxwell's introduction of probability theory into science was an important innovation. Previously probability theory had been much used in connection with dice and card games, particularly by the famous physicist, mathematician, and philosopher, Blaise Pascal (1623–1662). Maxwell saw the humour of the fact that he, a devout Presbyterian and therefore opposed to gambling, would be getting into this field. In a letter to an old school friend and future biographer, the classicist Lewis Campbell, he wrote 'The true Logic for this world is the Calculus of Probabilities. This branch of Math., which is generally thought to favour gambling, dicing and wagering, and therefore highly immoral, is the only "Mathematics for Practical Men".' This was a very perceptive comment, for as it turned out the theory of probability was just what was needed to understand the second law of thermodynamics—and many other things.

Maxwell's first derivation of the distribution law, although leading to the right mathematical expression, was not entirely satisfactory, and he had to grapple with the problem for some years before finding something with which he was happy. In

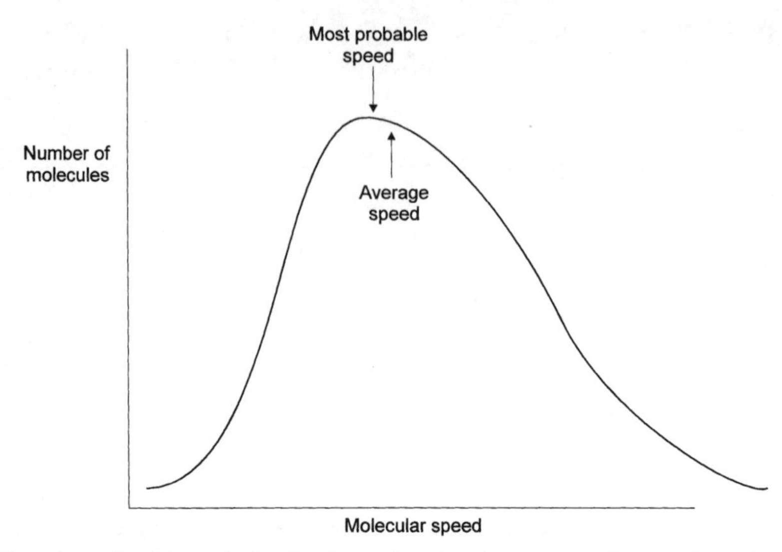

Fig. 16 The Maxwell distribution of molecular speeds. The number of molecules having speeds very close to a certain speed is plotted against the speed. Few molecules have a very low speed, and few a very high speed, but more have intermediate speeds. The maximum corresponds to the most probable speed, and the average speed is slightly larger than that.

1867, when he was in 'retirement' at Glenlair, he published a much improved version of his kinetic theory, including a better derivation of his distribution law. Maxwell's famous electromagnetic theory is often regarded as his most important achievement, since its consequences, for example in leading to radio transmission, have been so widespread. His theory of the distribution of molecular speeds is perhaps just as important. Particularly in the hands of Ludwig Boltzmann, of whom we shall hear much more in the next chapter, it led to the theory of the distribution of molecular energies, and to many more advances in our understanding of molecular systems.

Although he made important contributions to kinetic theory, especially by his distribution law, Maxwell always had serious doubts about its validity. The reason for this was that the theory seemed to be incapable of explaining the specific heats of gases. These specific heats are the amounts of energy required to raise the temperature by a particular amount. Maxwell was especially interested in the specific heats of gases that had been measured under two different conditions: with the gas held at constant pressure, and with it held at constant volume. He paid particular attention to the ratio of the specific heat obtained at constant pressure to that at constant volume, this ratio being given the symbol γ (Greek gamma).

In his 1857 paper Clausius had also worried over this problem and could not find a solution to it. In his 1860 paper Maxwell considered the problem in more detail with equal lack of success. The essence of the difficulty is as follows. An atom by itself has only one type of motion, translational motion, and we say that it has three degrees of translational freedom, by which we mean that all that it can do is to move about in three dimensions. A molecule like oxygen or nitrogen, on the other hand, contains two atoms (we call it diatomic) and in addition to its three degrees of translational motion it has some modes of vibrational and rotational motion. Maxwell's kinetic theory of gases, particularly his theory of the distribution of speeds, led him to the conclusion that the energy must distribute itself equally between the different degrees of freedom. This is referred to as the *principle of equipartition of energy*. We will not go into details here, because there are some complications, but Maxwell thought that the theory for diatomic molecules required γ to be 1.33. Experimentally, however, it is more like 1.4.

Because of discrepancies of this kind Maxwell became discouraged with the kinetic theory and ended his 1860 paper with the comment 'This result of the dynamical theory, being at variance with experiment, overturns the whole hypothesis, however satisfactory the other results may be.' He was, however, wrong in thinking that the whole hypothesis should be overturned; it just needed to be modified by the quantum theory, which of course had not yet been thought of. After that theory was developed, in the early years of the twentieth century, it became clear that the reason for the specific-heat discrepancy was that energy is quantized, not that kinetic theory is unsound. When, over 20 years after his death, the theory was suitably modified by the quantum theory, it gave a perfectly satisfactory interpretation of all the experimental results on specific heats. This will be further discussed in Chapter 7.

Arising from his kinetic theory Maxwell had another particularly important idea, the object of which was to explain the basis of the second law of thermodynamics. This was the concept of what has come to be called *Maxwell's demon*. This imaginary creature did not come to light in a scientific paper, as is usual with important scientific ideas, but was born in a letter that Maxwell wrote on 11 December 1867 to his old friend Tait, with whom we became acquainted in the last chapter. Throughout their adult lives the two men carried on a lively and amusing correspondence—sometimes, since they were thrifty Scots, by means of postcards which required only a halfpenny stamp instead of the penny stamp then needed for a letter. The point of this particular letter written by Maxwell was to show how, in principle (but almost never in practice), the second law could be violated.

Maxwell considered a vessel divided into two compartments A and B, separated by a partition which had a hole in it that could be opened or closed by 'a slide without mass' (Fig. 17). The gas in A was at a higher temperature than the gas in B. In a gas at a higher temperature, the average speed of the molecules will be greater than if the gas is cooler, but there is a distribution of speeds. Maxwell imagined 'a finite being', later called a demon, who knew the speeds of all the molecules. This creature could open the hole for an approaching molecule in A when its speed was low, and would allow a molecule from B to pass through the hole into A only when it was moving fast. As a result of this process, said Maxwell,

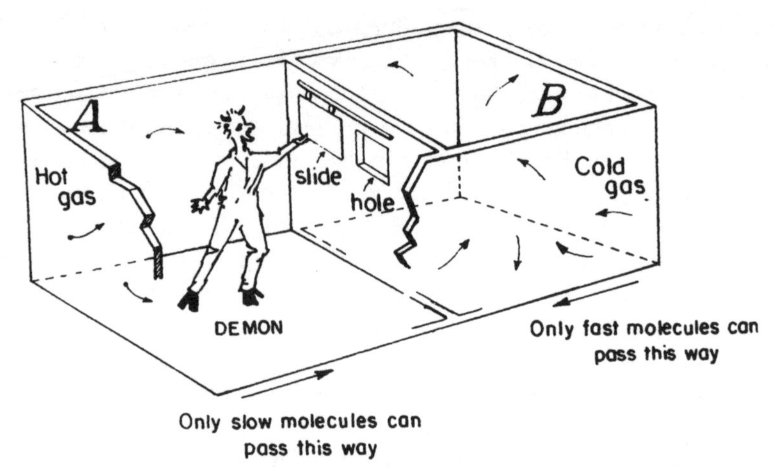

Fig. 17 Schematic representation of Maxwell's demon. The left-hand compartment A contains gas that is hotter than the gas in B, which means that the molecules in A are on the average moving faster than the molecules in B. The demon allows only fast-moving molecules to move from right to left, and only slow molecules to move from left to right. The hot gas therefore gets hotter, and the warm gas cooler. Maxwell used this imaginary device to explain the second law of thermodynamics, not to show that it is invalid, which is sometimes thought.

'the hot system has got hotter and the cold colder and yet no work has been done.'

Of course, Maxwell did not imagine that his 'finite being' could exist; he emphasized that his intention in inventing it had simply been to provide us with an understanding of why the second law of thermodynamics applies; it is just a matter of probability. In spite of clear statements by Maxwell, his demon has been widely misunderstood. Many students, and even some experts like the distinguished philosopher Sir Karl Popper (1902–1994), have believed that Maxwell thought he had shown the second law to be invalid. Maxwell emphasized, however, that he regarded his demon as just a device to help people to understand the second law. He made this clear, for example, not only in the sentence quoted at the head of this chapter but in a statement he made to Tait in the form of a humorous catechism, 'Concerning demons', of which item 3 read:

> 'What was their chief end?'
> 'To show that the second law has only a statistical certainty.'

It is interesting to note that earlier discussions of the second law made no reference to a process that today is often used as a good example of the application of the law, namely the mixing of two different gases at constant temperature. Maxwell's demon could reverse such a process; air, for example, could be separated into oxygen and nitrogen by the use of a demon who would allow only oxygen molecules to pass in one direction and only nitrogen molecules to pass in the other. The earlier workers thought of the second law as relating only to engines, where there is a passage of heat from a higher temperature to a lower one, and a mixing process is not obviously related to such transfers of heat.

One particular point about entropy and the second law requires a little discussion. It relates to the role of information in connection with the second law, and it clearly relates to what has come to be called Gibbs's paradox. It was explained by Gibbs in the third of three papers he published between 1873 and 1878. We may imagine a vessel divided into two halves separated by a partition. Suppose first that the two compartments contain two different gases at the same pressure and temperature. If the partition is then removed there will be an increase of entropy because the gases have mixed, and the value of the entropy is easily calculated. The increase in entropy is due to the increase in randomness or disorder.

Gibbs compared this with the situation if the two gases instead of being different are identical. There is then obviously no increase of entropy when the partition is removed, because nothing in effect has happened; there is no longer any mixing of gases. This difference sometimes causes consternation, because from one point of view what has happened in the two cases is really the same. Gibbs himself did not regard the difference in the two entropy changes as really paradoxical. He saw clearly that the difference depends on the information we have about whether the gases are the same or not. Obviously in computing an entropy change we have to take into account certain information.

The same point arises if we are considering the entropy change in the shuffling of 52 playing cards. We can say nothing about the entropy change unless we have further information. If by mistake the cards have been printed so that all are identical, there will be no entropy change on shuffling. If, however, the cards have been printed normally, there is an increase in entropy when an ordered deck is shuffled. We call this an *informational entropy*, arising from the fact that for an ordered deck we accept only one arrangement of cards out of the vast number of possible arrangements. Probability theory leads to the result that the possible number of arrangements in a deck of 52 is about 4.4×10^{66}. When we mix two gases or liquids, of course, the informational entropy is much larger than for the shuffling of cards, because the number of molecules in a gas of visible size is much greater than the number of cards. For example, suppose that we mix a litre of ordinary water with a litre of heavy water, D_2O. The chance that the mixture would spontaneously separate into the two separate liquids works out to be 10 raised to the power of 10^{23}, that is one followed by 10^{23} zeros. We will consider these enormous numbers again in the next chapter.

This informational entropy can also be called the *conformational entropy*, because it depends on the number of arrangements, or conformations, of the cards or molecules that we are dealing with.

This matter of informational contributions to entropy is of great importance, but is widely misunderstood. A proper understanding of it has formed the foundation of modern information theory. Sometimes the fundamentals of this subject are presented with reference to the functioning of Maxwell's demon. The only way that the demon could reverse the ordinary course of events and cause a warm gas to get warmer and a cold one to get colder when the two gases are side by side (as in Fig. 17) is by acquiring information about the speeds of the individual molecules. If this occurred, and the gases did incur a decrease in entropy, this would be overcome by the entropy change that is involved in the use of information.

Six

Chance and the distribution of energy

Entropy is the Arrow of Time.

Arthur Eddington, *The Nature of the Physical World*, 1927

Much further work on the relationship between entropy and probability, and the distribution of molecular energies, was also carried out by the Austrian physicist Ludwig Boltzmann (1844–1906; Fig. 18). Some of his struggles associated with this problem are the subject of the present chapter. The matter is somewhat complex and confusing, and it is best to summarize the final outcome at once. As Maxwell had clearly recognized with his demon, the second law is all a matter of chance, or probability. Boltzmann's work showed this to be true, but unfortunately he somewhat clouded the issue by suggesting at times that the law could be derived without taking chance into account.

Fig. 18 Ludwig Boltzmann (1844–1906), who developed a kinetic theory of the approach to equilibrium, and whose work led to the important new branch of science called statistical mechanics.

Boltzmann was born in Vienna and attended the University of Vienna. He received his doctorate in 1866 and continued there for a year as a research assistant. In 1867 he was appointed an assistant professor at the university. In 1869 he became professor of mathematical physics at the University of Graz. Boltzmann had a pathologically unhappy and discontented disposition, as is suggested by the number of moves he made during his career. In 1873 he became professor of mathematics at the University of Vienna, but soon realized that he did not enjoy teaching mathematics and in 1876 returned to the University of Graz as professor of experimental physics—rather inappropriately, as his work was entirely theoretical. From 1889 to 1893 he was professor of theoretical physics at the University of Munich. In 1894 he returned to the University of Vienna as professor of theoretical physics but in 1900, being unhappy there, he accepted a professorship at the University of Leipzig. He soon decided, however, that he preferred Vienna after all and returned there in 1902 to succeed himself as professor of theoretical physics.

This brief summary of Boltzmann's academic career gives us some idea of his discontent and restlessness, but it tells only a small part of the story. None of the career moves he made proceeded smoothly, but each was accompanied by much hesitation and indecision; positions were accepted but then rejected, and resignations were tendered and soon repudiated. This eccentric behaviour applied also to his personal life. For many years he was a manic depressive, passing through many periods of acute unhappiness. His condition became markedly worse in about 1888 before his move to the University of Munich. During his later years at the University of Vienna he became incapable of lecturing or carrying out any of his academic duties, made several attempts to take his own life, spent some time as a voluntary patient in a mental hospital, and finally committed suicide.

In considering his scientific work it is useful to bear these facts in mind, as they help us to appreciate some of the details of his scientific accomplishments. There is no doubt that he was one of the most distinguished of scientists and that he made pioneering advances of the greatest importance. At the same time he made some serious mistakes, which he sometimes corrected, but some of which persisted, and about some of which he vacillated. Unlike Maxwell, who wrote with great clarity, Boltzmann was wordy and often confusing. As a result he was constantly involved in controversy, sometimes because he was wrong and sometimes because he had been misinterpreted as a result of his obscurity.

Much has been written about Boltzmann's manic-depressive syndrome, and no one can give a satisfactory explanation for it. It no doubt had a largely genetic basis, and psychiatrists today could probably have controlled it by medication. A contributing factor may have been that Boltzmann was more philosophically inclined than most scientists, who tend to accept the world as they find it; Boltzmann, on the other hand, was highly introspective and inclined to brood about his scientific findings as well as about his personal life. The fact that he was brought up a Roman Catholic may have created spiritual difficulties for him, since at the time his Church was quite hostile to scientific advances; in 1864 Pope Pius

IX issued a *Syllabus of Errors* which was a blanket condemnation of the theory, methods, conclusions, and practice of science. Boltzmann was a strong supporter of Darwin's theory of evolution, which the Church then specifically condemned and officially (although not in practice) continued to do until 1996. Some of his philosophical writings show clearly that his work had led him to the pantheistic belief that God is no more than a manifestation of the laws of nature, a view that his Church would have thought heretical. There seems never to have been any confrontation between him and the Church authorities, perhaps because they did not read his papers—or, if they did read them, they did not understand them. All his life Boltzmann remained a Roman Catholic, at least nominally, and it seems likely that this discrepancy between the philosophical opinions to which his science had led him, and the philosophy taught at the time by his Church, troubled him deeply.

Boltzmann began to make contributions to kinetic theory while still a student at the University of Vienna, his interest having been inspired by two of his teachers, Josef Stefan (1835–1893) and Johann Josef Loschmidt (1821–1895), both of them men of great scientific distinction. Loschmidt was particularly interested in atomic theory and the kinetic theory of gases; he is famous for obtaining the first reliable estimate of atomic sizes. Boltzmann's first paper of any significance, published in 1866 when he was 22 and a year before he obtained his doctorate, was entitled 'On the mechanical meaning of the second law of thermodynamics'. He started by commenting that the first law had been known for a long time, and a man of 22 does tend to think that about 20 years is a long time. He noted that the second law could only be established by 'roundabout and uncertain methods', and said that his object was to give a 'purely analytical, completely general, proof of the second law of thermodynamics, as well as to discover the theorem in mechanics that corresponds to it.' In other words, he thought that he could arrive at the second law without taking probability into account. This is a mistaken view, and one that Boltzmann never seemed to throw aside entirely; as we shall see, in later papers he made contradictory statements on the matter.

It is interesting to note that Clausius also tried to explain the law as a necessary consequence of the principles of mechanics. In 1870 and 1871 he published papers in which he too claimed to have provided a mechanical explanation of the second law, the title of one of his papers being 'On the reduction of the second law of thermodynamics to general mechanical principles'. This paper was at once followed by one from Boltzmann, then at the University of Graz, pointing out that his 1866 paper included essentially the same treatment as that given by Clausius, and concluding with the rather sardonic comment:

> I can only express pleasure that an authority with Herr Clausius's reputation is helping to spread the knowledge of my work in thermodynamics.

This comment, which is typical of Boltzmann's unsophisticated and clumsy style, probably did not endear him to Clausius, already a distinguished professor and

Boltzmann's senior by 22 years. However, in a paper published in 1872 Clausius graciously conceded Boltzmann's claim, explaining that extraordinary demands on his time had made it difficult for him to follow the scientific literature. In any case, of course, these attempts to circumvent the theory of probability were misguided. For reasons that are difficult to understand, Clausius never made much use of Maxwell's statistical interpretation of the second law, although he continued to develop kinetic theory. Boltzmann's later attitude, however, was that he sometimes accepted the probability interpretation but often wrote otherwise.

Letters written by Maxwell to his friend Tait after these mechanical theories had been put forward showed that Maxwell regarded them with some amusement, realizing that Clausius and Boltzmann had missed the point. For example, in a letter to Tait dated 1 December 1873, Maxwell commented that 'It is rare sport to see these learned Germans contending for the priority of the discovery' that the second law of thermodynamics is no more than a consequence of the laws of mechanics, with no reference to the laws of probability. Maxwell never attacked Clausius and Boltzmann directly, but made his own position clear in his *Theory of Heat.* It seems, rather surprisingly since their work was so closely related, that Maxwell and Boltzmann never met or even corresponded.

In 1868, two years after his unsuccessful attempt to explain the second law on purely mechanical grounds, Boltzmann published a lengthy paper of great importance in which he extended Maxwell's theory and treated the distribution of energy. In his 1866 paper he had made no mention of Maxwell's work, but he was then learning English, particularly with a view to studying Maxwell's papers on the electromagnetic theory, and by the time he wrote his 1868 paper he had also become familiar with Maxwell's theory of the distribution of velocities. As noted in Chapter 5, Maxwell's derivation of the distribution equation, first published in 1860 and modified in 1867, was somewhat abstract, and in his 1868 paper Boltzmann produced a more convincing derivation. He considered a column of gas, such as the Earth's atmosphere, for which the decrease of pressure with increasing height above sea level could be explained by the decrease of the gravitational energy. An atom moving upwards in a gravitational field behaves like a ball thrown upwards, its changing velocity and energy varying in accordance with Newton's mechanics. On this basis Boltzmann obtained an equation for the variation of pressure. He then went further and considered the situation in which a potential energy is involved, and then deduced an important expression for the fraction of molecules having energy E. He found that this fraction is proportional to the fraction

$$\exp(-E/k_{B}T)$$

where k_B is now called the *Boltzmann constant* and T is the absolute (kelvin) temperature (see the note on p. xii for an explanation of this type of expression). Since $-E/k_{B}T$ is a negative quantity, $\exp(-E/k_{B}T)$ is bound to be a fraction. The great importance of Boltzmann's derivation is that he had shown that the fraction $\exp(-E/k_{B}T)$ not only applies to the energy of a molecule that is due to its motion from one place to another (its kinetic energy of translation as we call it), but also applies to all kinds of energy, such as energy due to a gravitational field and internal energy of vibration.

An important consequence of the *Boltzmann factor* $\exp(-E/k_BT)$ is that it leads to the useful conclusion that the average energy of a molecule at temperature T will be something like k_BT, depending to some extent on the nature of the molecule. It is interesting that Boltzmann himself expressed this factor in a different form, one that did not involve the Boltzmann constant. We shall see in the next chapter that it was Max Planck who first used the Boltzmann constant, in 1900 when he first formulated his quantum theory. In his paper on the subject Planck also gave for the first time a numerical value for the constant.

One feature of Boltzmann's derivation is that he regarded the total amount of energy as distributed among molecules in such a manner that all ways of doing so are equally probable. This follows from the principle of equipartition of energy, which had earlier been derived by Maxwell (see Chapter 5). Entirely for mathematical convenience, Boltzmann regarded the energy as composed of a large number of tiny packets of energy. The expression he finally obtained reduced to Maxwell's velocity distribution expression if the packets of energy were taken to be infinitesimally small. This approach is interesting since it foreshadowed Planck's quantum theory, according to which there really are small packets or quanta of energy. In fact, Planck made good use of many of Boltzmann's mathematical procedures in developing his quantum theory, as we shall see in the next chapter.

Boltzmann made these important contributions, and a few others, while still not yet 25; by that time he had published a total of eight papers and already his work was beginning to attract attention. Although Boltzmann never had any direct contact with Maxwell, his work on the distribution of energy led to a curious indirect contact, through Boltzmann's former professor Loschmidt who received a letter from Maxwell commenting on the 'outstanding work of your student'. Maxwell had mistakenly thought the work to have been done under Loschmidt's direction. This remark in Maxwell's letter greatly pleased Boltzmann, who enjoyed quoting it from time to time.

The next important problem he dealt with was the way in which systems approach a state of equilibrium, his major contribution on this subject appearing in 1872, when he was professor of mathematical physics at the University of Graz. Approach to equilibrium had been considered from two different points of view: in terms of the dissipation of energy, following Kelvin, or better in terms of an increase of entropy, as was done by Clausius. Boltzmann's purpose in this 1872 paper was to show, in terms of pure kinetic theory without bringing in probability, how a system behaves as it approaches a state of equilibrium, the entropy steadily increasing as it does so. By going through a rather lengthy mathematical treatment of kinetic theory he was led to a function, now given the symbol H, which involves among many other terms the Boltzmann factor $\exp(-E/k_BT)$. A simple way of thinking about it is to note that with its sign changed it is proportional to the entropy; it is a measure of the negative of the entropy, which is sometimes called *negentropy*. We recall from Chapter 4 that Clausius showed, in his formulation of the second law, that the total entropy is bound to increase as time goes on. It follows, since H is proportional to the negative of the entropy, that the value of H for any system plus its environment is bound to *decrease* when any spontaneous process occurs.

The significance of the function H is that it extends the definition of entropy to include states that are not at equilibrium, in contrast to pure thermodynamics which only deals with systems at equilibrium. In other words, it is a theory of the dynamic approach to equilibrium. The theorem was originally referred to as 'Boltzmann's minimum theorem', but is now usually known as 'Boltzmann's H theorem'. The reason that the letter H is used is trivial but interesting—it was a mistake in transcription. Boltzmann had himself used the letter E, but a British scientist who saw that letter in German script mistook it for an H, which somehow stuck.

Soon afterwards Boltzmann dealt with an objection that had been raised to his theorem, the so-called 'reversibility paradox' which had been suggested by Kelvin in a paper that appeared in 1874 and again by Boltzmann's friend and former professor, Josef Loschmidt, two years later. The essence of the objection is as follows. Suppose that we first consider the passage of a system such as a gas from a state at which it is not at equilibrium to the state of equilibrium; according to Boltzmann's treatment H must be decreasing. Now there could be no objection to the existence of a state in which all the molecular motions are reversed, when the process would be occurring in the opposite direction, away from equilibrium, with H now increasing. How then can Boltzmann insist, as a matter of mechanics, that H must always decrease? Boltzmann's response to this, in a paper that appeared in 1877, was that there is a vast number of possible equilibrium states, but if we specify a particular initial state, we are confining ourselves to just that one state. As a matter of probability, therefore, motion towards equilibrium is vastly more likely than motion in the opposite direction. He admitted that we could chose states that would move away from equilibrium with an increase in H and a decrease in entropy. However, such a situation would be highly unlikely. This argument of course involves probabilities, and contradicts what Boltzmann had implied in his 1878 article, namely that the decrease in H follows from pure kinetic theory and that there was no need to invoke probability theory. It is clear that we cannot avoid probability theory, as Maxwell had maintained all along. Much confusion resulted from the fact that for many years Boltzmann was inconsistent in his statements about the matter, sometimes using the probability argument and sometimes saying that it was not necessary to invoke probability theory.

A related problem, referred to as the 'recurrence paradox', arose a good deal later from a theorem in mechanics first propounded in 1893 by the French physicist and philosopher Jules Henri Poincaré (1854–1912). According to this theorem, any system such as a gas must, in the course of time, eventually reach every configuration possible. Poincaré, and later and more persistently the German mathematician and physicist Friedrich Ferdinand Zermelo (1871–1953)—whom

Boltzmann often referred to as *dieser Helunke* (that rascal)—argued that therefore the H theorem cannot always be valid, since a system can pass to any other state. Boltzmann's answer to this challenge, published in 1896, was similar to his answer to the reversibility paradox. On the molecular scale the equilibrium state consists of a vast number of possible configurations. It is true that in principle a particular chosen state can recur if we wait long enough, but we would have to wait an immensely long time. Thus, although the H theorem could in principle be violated, such a violation is highly unlikely. Obviously he was again invoking probability to justify the H theorem.

To emphasize this point Boltzmann made an estimate of the average time it would take for a gas containing 1 million million million (10^{18}) atoms, contained in a volume of 1 cubic centimetre at room temperature, to start at one precise dynamical state and for exactly the same state to recur. He found that the number of seconds it would take would be one followed by many trillions of zeros. To give some idea of the enormity of this number he pointed out that if every star in the sky had the same number of planets as the Sun, and if every one of these planets had the same population as the Earth, and if every one of these people lived a trillion (10^{12}) years, the number of seconds in their combined lifetimes would be much smaller than the enormous number he had calculated for the gas. The possibility of a violation of the second law is therefore so remote that for all practical purposes we can forget about it. As to Zermelo's argument, Boltzmann said that suppose we had a thousand dice and tried to throw them so that all would come up as ones. Such, he pointed out, was fantastically unlikely. Zermelo, he commented rather sarcastically, 'is like a dice player who concludes that something is wrong with the dice because such an occurrence has not presented itself to him.'

Thus, when pressed, Boltzmann again invoked the probability argument to justify his argument. Sometimes, however, he insisted that he had arrived at his function H on the basis of pure mechanics. He seems never to have properly understood the apparent paradox, which is a rather subtle one. The function H was indeed derived by Boltzmann from pure mechanics, without explicitly involving probability, and it does decrease with time, but how is this possible? The answer is that in his derivation of the expression for H he had made some hidden assumptions about the characteristics of typical collisions between atoms. He had, as a convenient approximation, neglected unusual types of collisions, and by doing so had unwittingly 'loaded the dice'; he had left out the terms that would cause H to lead to the wrong conclusion, and had included only the ones that led to the second law. His H theorem had not done what he had originally expected and intended: lead to the second law by a purely dynamical argument. That does not mean that it was a waste of time. It does allow us, for example, to gain some appreciation of the approach to equilibrium. The kinetic theory he developed was useful in leading ultimately to the understanding that the second law is no more than a matter of chance, but a chance that was to all intents and purposes a certainty. The irony is that Maxwell had understood this all along; no wonder he felt a little superior when he saw all the effort going into trying to avoid probability theory. He knew that the H theorem only seemed to offer certainty, when in fact there was only an extreme improbability that the second law could be violated.

From 1871 to 1887 Boltzmann published several papers of great importance in which he showed how his factor $\exp(-E/k_B T)$ allowed all of the properties of a system to be calculated. This is what the important field of *statistical mechanics* is mainly concerned with, and the methods Boltzmann used are essentially those commonly used today. What he did in essence was to consider the distribution of energy in terms of putting atoms into pigeon holes corresponding to different values of the energy. The total amount of energy has to be the same for each arrangement, but this condition can be satisfied in a vast number of different ways. Arguing in this way he proceeded to show how to calculate the number of equivalent ways that the atoms could be fitted into the compartments (subject to the condition that the total energy must be the same). This number he equated to the probability of the atomic distributions. He showed that the most likely distribution was the one given by the Maxwell distribution equation.

He went even further, by pointing out that the closer that any distribution was to the equilibrium one, the more likely it was. The farther away it was, the less likely it was. By reasoning in this way Boltzmann was led to his famous relationship[1] between entropy and probability, W:

$$S = k_B \log W.$$

In this equation, which is engraved on Boltzmann's tombstone in Vienna, W is the number of possible ways of making a given distribution of atoms or molecules, corresponding to a given total energy of the system. These molecular configurations are now referred to as *microstates*. This equation allows an expression for the entropy to be obtained from the statistical distribution, and from this expression the other thermodynamic properties can be calculated.

Note that in this relationship Boltzmann had actually introduced an important extension to the Clausius definition of entropy. We saw in the last chapter, with reference to the Gibbs paradox (p. 57), that when considering an entropy change it is important to include the informational entropy contributions. For example, in dealing with the entropy change when two gases are brought together, we get a different answer if we use the information that the two gases are different from what we get if we think they are identical. If we proceed by using Clausius's definition we have to add the informational entropy rather arbitrarily. The Boltzmann formulation, on the other hand, is more satisfactory in that it deals with the informational contribution quite naturally.

It is helpful to consider some examples, and we will start with the shuffling of a deck of 52 cards. We noted in the last chapter, in connection with the Gibbs

[1] See the note on p. xii for an explanation of logarithms. The logarithm used here is not the common logarithm but the natural logarithm, which is the power to which the number e (= 2.718 28 ...) has to be raised to get the number in question. Today we usually write the natural logarithm as ln W.

paradox, that if by mistake the cards have been printed so that all are identical, there will be no entropy change on shuffling. If, however, the cards have been printed normally, there is an increase in informational entropy when an ordered deck is shuffled, arising from the fact that for an ordered deck we accept only one arrangement of cards out of the vast number of possible arrangements. For a deck of cards probability theory leads to the result that the possible number of arrangements is about 4.4×10^{66}. It is therefore fantastically unlikely—one in 4.4×10^{66}—that a single shuffling will produce an ordered deck. We can see how unlikely it is by imagining that one could shuffle cards at the rate of one shuffle per second, and had been doing that day and night since the beginning of time, which is something like 10^{18} seconds ago. To have a sporting chance of getting an ordered deck one would instead have to shuffle for 10^{48} times as long as that. If in real life someone shuffles a deck and gets an ordered one we are justified in saying that some trickery was involved.

If we mix two gases or two liquids, of course, the informational entropy is much larger than for the shuffling of cards, because the number of molecules in a gas of visible size is much greater than the number of cards. A litre of water contains roughly 10^{25} molecules. Suppose that we have a litre of ordinary water mixed with a litre of heavy water; what is the chance that it would spontaneously separate into ordinary water at one side of the bottle and heavy water on the other? The answer is 10 raised to the power of 10^{23}. It is impossible to visualize the magnitude of such a number, and the following may help a little. Suppose that I wanted to write this number down for you on a strip of paper as one followed by 10^{23} zeros. To save paper I will make the zeros only 1 millimetre wide, so that the length of the string of zeros will be 10^{20} metres or 10^{17} kilometres. The distance to the Moon and back is roughly 10^{6} kilometres, so that our string of zeros will go to the Moon and back 10^{11} times, or a hundred billion times. It would go the nearest star, Proxima Centauri, and back about a thousand times. I doubt whether I could afford so much paper.

All of Boltzmann's work depended on the real existence of atoms. Today we take atoms and molecules for granted, as there is now much direct evidence for them. In Boltzmann's time, although the evidence for atoms was largely circumstantial, the existence of atoms had become generally accepted among scientists. Most chemists at the time considered atoms to be real, and were devoting much effort to discovering how atoms join together to form molecules. For example, they would have argued, following John Dalton in the early years of the nineteenth century, that the fact that a substance like water always has the same composition, and can be represented by the formula H_2O, can only be explained on the hypothesis that molecules of water exist, and that each consists of one atom of oxygen and two of hydrogen. However, during the last decade of the century several influential voices were raised against the real existence of atoms, notably those of the Austrian physicist and philosopher Ernst Mach (1836–1916) and the German chemist Wilhelm Ostwald (1853–1932; Fig. 19). Their reasons, however, were different. Mach, whose name is still remembered in connection with the

Mach numbers for high-speed flight, was a leader of the positivist school of thought. According to him, concepts such as the atomic theory should not be accepted since the evidence for them must always be indirect or circumstantial, based on inference. The atomic theory might be helpful for classifying information, but he believed that one should not conclude that atoms really exist.

Ostwald thought that everything could be explained in terms of energy, a view with which Mach did not agree. Ostwald was so enthusiastic about his theory of energetics that he gave the name *Energie* to the Saxony estate to which he retired in 1906. Boltzmann had pointed out to Ostwald on a number of occasions, and especially at a meeting in the Baltic port of Lübeck in 1895, the fallacy of this argument. Ostwald would have argued, for example, that water always has the H_2O composition because for energetic reasons oxygen and hydrogen are bound to combine in that way. He did not say what the energetic reasons are, and failed to see that it is the existence of atoms that requires them to do so.

Ostwald remained sceptical of the real existence of atoms until 1909, when he was finally convinced by more direct evidence for them. Until that time his very influential textbooks of physical chemistry had given no explanations in terms of atoms. Mach never conceded the existence of atoms.

Fig. 19 Friedrich Wilhelm Ostwald (1853–1932), who played a key role in the development of the science of physical chemistry. He lucidly explained the subject in his well-known textbooks, and made important contributions in a wide range of topics. He was awarded the 1909 Nobel Prize in Chemistry.

It has sometimes been suggested that Boltzmann's personal unhappiness, culminating in several suicide attempts and his eventual successful suicide in 1906, was due in part to strain brought on by the many debates he had about the existence of atoms. It is true that Boltzmann could have considered his scientific efforts to have been wasted if atoms do not exist, because his theories depended critically on atoms. It is also true that Boltzmann in his later years became largely deserted by German scientists on this issue. At the same time he received much support from a number of distinguished scientists, especially in Britain. He received many invitations from abroad; in 1894, for example, he attended by invitation a meeting in Oxford of the BAAS, where his kinetic theory was discussed much more sympathetically than at some other meetings. On that same visit to England he was awarded an honorary degree of Doctor of Civil Law (DCL) by Oxford University. He was, incidentally, somewhat puzzled at not receiving a Doctor of Science degree, but an English friend explained to him that it must have been because of his work on the *laws* of thermodynamics. In 1905, a year before his death, he made a highly successful tour of the United States and again received an honorary degree. According to his own highly amusing and cynical account of his trip, entitled 'A German professor's journey into Eldorado', he greatly enjoyed himself. All in all, it would seem that his scientific career should have caused him great satisfaction rather than any distress.

It seems equally unlikely that his unhappiness was caused by his home life or his academic career. He always had many friends who gave him much support. In 1876, at the age of 32, he married Henriette von Aigentler and the union seems to have been a happy one. His wife was always supportive, and they had five children, two boys and three girls, to whom he was devoted. His academic career was successful; at the University of Graz he became Dean of the Faculty in 1878 and Rector (equivalent to President or Vice-Chancellor) in 1887. It is of interest that Ostwald, perhaps with Boltzmann in mind, had given much thought to human happiness. He classified persons of genius into two broad types, classicists and romanticists. The classicists are phlegmatic, melancholic, and react slowly; the romanticists (in which group he included himself) are choleric, and react precipitately. He even devised a formula for happiness, for which he used the symbol G, for *Glück*. According to him G is proportional to

$$(A - W)(A + W) = A^2 - W^2.$$

Here A is the energy spent on doing things that one enjoys, and W is the energy spent on doing things that are disagreeable. This formula does seem rather naïve and unrealistic; we all know people who can be happy doing things that are unpleasant, and Boltzmann himself was a good example of someone who spent his life doing what he wanted to do but became unhappy about it. For him one must look for a deeper cause, an inborn pathological condition.

The manner of Boltzmann's suicide was particularly tragic. By 1906, when he was 62, he had become so despondent that he was unable to lecture or perform other academic duties; the authorities of the university had officially acknowledged his condition. To aid his recovery he went with his wife and daughters for three weeks to the coastal village of Duino, near Trieste, and at the end of the holiday he had seemed much improved. About six in the evening of 5 September his wife and daughters went down to the sea to bathe, leaving him behind in the hotel. He had said that he would join them later, but never came. Finally their 15-year-old daughter Elsa returned to the hotel, and to her unspeakable horror found him hanging from the crossbar of a window. For the rest of her life she could never speak of her terrible discovery. Since Boltzmann had always been an affectionate father, and his youngest daughter Elsa had been the apple of his eye, it tells us something of the extent to which he was emotionally disturbed that he would commit suicide in such a manner that there was a good possibility that Elsa or any others of his family would find his body.

After a Roman Catholic funeral Boltzmann was buried in a cemetery on the edge of Vienna, in a special section containing *Ehrengraben* (honour graves) where Beethoven, Schubert, Brahms, and other famous people are buried. For a time the grave fell into neglect but in 1922 his coffin was removed from it. This proved troublesome since another person had been buried above him and the family would not allow their relation to be disturbed. A sloping shaft was therefore dug, and Boltzmann was reburied in the Central Cemetery of Vienna. In 1933 a monument was erected bearing his famous formula that relates entropy to probability.

This chapter has told a tangled tale, of doubts and controversies, about something that in essence is quite straightforward. The question at issue, which some people found philosophically troublesome, was whether a law of nature, which the second law of thermodynamics certainly was, could just be a matter of chance. Every other one of the laws of nature that had previously been deduced was quite definite and straightforward. The second law was the only law that had a different basis; it was not absolutely true but was only approximately true, as a matter of chance. It was true that the chance of the law being violated was usually exceedingly remote. Some scientists, however, found it philosophically unsatisfactory that what was supposed to be a law of nature was true only by chance.

In addition, to some there was uneasiness on religious grounds. If the second law of thermodynamics were true only as a matter of chance, where was the hand of God? It is interesting that Maxwell and Kelvin, both brought up as Scottish Presbyterians and both devout at any rate in a formal sense, reacted differently to this argument. Kelvin was troubled by it, but Maxwell was not. He more than Kelvin followed in the tradition laid down 200 years earlier at the founding of the Royal Society. The Charter of the Society stated unambiguously that science and religion were to be kept completely separate. This estimable tradition was clearly expressed in 1860 by the Revd Frederick Temple (1821–1902), who later became Archbishop of Canterbury. His sermon, preached at the meeting in Oxford of the BAAS, included the passage:

> The student of science now feels himself bound by the interests of truth, and can admit no other obligation. And if he be a religious man, he believes that both books, the book of nature and the Book of Revelation, alike come from God, and that he has no more right to refuse what he finds in one than what he finds in the other.

Maxwell, and the majority of British scientists of the time, followed in this tradition. To them, kinetic theory and the theory of chance still left room for a Creator. Today the majority of scientists are agnostics, but it is not usually because of the findings of science. It is because their devotion to the scientific method leads them to reject authoritative religious writings, and to come to their own conclusions.

Seven

Packets of energy

> A new epoch was inaugurated in physical science by Planck's discovery of the quantum of action.
>
> Niels Bohr, in *Philosophy in Mid Century*, 1958

By the beginning of the final decade of the nineteenth century most scientists felt satisfied that they largely understood the workings of nature. The laws of thermodynamics and other laws of nature had become well understood, and Maxwell had formulated a beautiful theory which interpreted the wave nature of light. Then, within a matter of a few years, the situation changed completely. X-rays were discovered by Wilhelm Konrad Röntgen (1845–1923) in 1895, and in 1897 Antoine Henri Becquerel (1825–1908) discovered radioactivity. In the same year the Cambridge physicist Joseph John Thomson (1856–1940) gained some detailed understanding of the properties of the electron. In July 1898 Marie Curie (1867–1934) and her husband Pierre Curie (1859–1906) discovered the radioactive element polonium, and in December of the same year they discovered the more radioactive element radium. All of these achievements completely transformed both science and society.

For a few years scientists again felt that after these exciting discoveries there could hardly be much more to learn. How wrong they were! Little did anyone guess that within only five years from the beginning of the new century there would be two more theories that would bring about radical changes in the way we think about science, and which were to have great practical consequences. These two theories were the quantum theory, introduced by Max Planck in 1900 and in a more comprehensive form by Albert Einstein in 1905, and the theory of relativity, due entirely to Einstein in 1905 and later. The present chapter gives some idea of the quantum theory, while the next deals with relativity and with some of its consequences. Both of these theories are important for the deeper understanding of energy.

Until the birth of the quantum theory it had been taken for granted that energy, including the energy of radiation, is continuous; in other words, that one could think of an atom or molecule as having any amount of energy. The essence of the quantum theory is that this is not the case; instead energy comes in small packets, or *quanta* (from the Latin *quantum*, how much?). According to the quantum theory we cannot have any amount of energy we demand—just as in a shop we are not allowed to buy any amount of milk we like but have to buy a carton, bag, or bottle of it, and we are not allowed to pay for anything in fractions of the smallest coin. What is special about the energy situation is that the packets are so incredibly small that it is hard to know that they exist. For example, when we drive a car it is impossible for us to be able to tell that the speed at which we drive is quantized, and that only certain speeds are allowed to us. The permitted speeds are so fantastically close together that the most careful measurements would never detect that there is any quantization. In view of this it is amusing that journalists, politicians, and others are fond of talking about 'quantum leaps' when they think they are making an important advance. They do not realize that these 'leaps' are quite undetectable in our everyday lives. The next time you hear a politician promising a quantum leap, remember that they are tiny and occur with completely unpredictable results.

To understand the quantum theory we need to know something about vibrators or oscillators, and we can consider a weight suspended at the end of a spring. We will start with it motionless, as shown in Fig. 20(a). If we pull it down to a position A as shown in Fig. 20(b), and then release it, the weight will then vibrate up and down between the positions A and B. The distance between A and B is known as the *amplitude* of the vibrations. Another important characteristic of the vibration is the time that it takes for the weight to move from position A up to B and back again to A; this time is called the *period* of the vibration. The number unity divided by the period (which we call the reciprocal of the period) is called

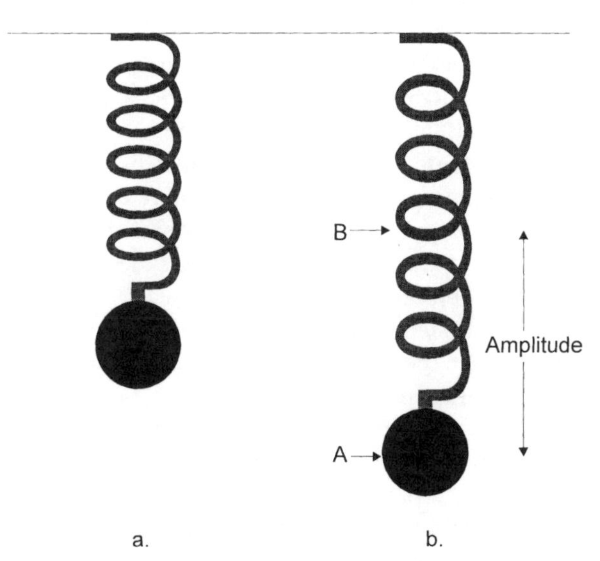

Fig. 20 (a) A weight attached to a spring, in its undisturbed state. (b) The weight has been pulled down to a position A. When it is released the spring pulls it back to the position B, after which it oscillates between the two positions. By definition the amplitude of the vibration is the distance between A and B. The time that it takes to travel from A to B and back is called the period of the vibration, and the reciprocal of that time is the frequency of the vibration.

the *frequency* of the vibration. Suppose, for example, that it took 2 seconds for the weight to move from A up to B and back to A: we would say that the period was 2 seconds and that the frequency was one divided by two or 0.5 per second, which we can call 0.5 reciprocal seconds.

Similar laws of motion arise in other oscillating systems that we meet in scientific work. A pendulum, such as a clock's pendulum, and a child's swing behave in the same way. They too undergo a to-and-fro motion which obeys much the same laws as the weight hanging from the spring. The atoms in a solid also are constantly undergoing a vibrational motion and they are often referred to as oscillators. Our air is composed mostly of nitrogen and oxygen, and nitrogen and oxygen molecules are both diatomic, by which we mean that each of the molecules consists of two atoms held together by chemical bonds which have the same characteristics as the spring shown in Fig. 20. We can represent the molecules as

N ~~~~ N and O ~~~~ O

and can think of them as continuously undergoing vibrations in which the bonds alternately shorten and lengthen.

One important characteristic of these oscillating systems is that to a good approximation the frequency remains the same whatever the amplitude of the vibration. Suppose, for example, that we pulled the weight in Fig. 20(a) down to a lower position than A, which means that we would be using more energy and would be stretching the spring more. After we released the weight it would rise to a higher position than B, which means that the amplitude of the vibration is larger than it was before. In spite of this, however, the frequency (and therefore the period) remain almost exactly the same. When there is a greater amplitude the weight travels a greater distance between the extremities of each vibration. It does so in the same period of time, which means that it has to move faster on the average to make the longer trip, and it thus has more energy. Vibrations with a greater amplitude are thus more energetic, but have much the same frequency of vibration. Musicians, of course, know this; a pianist can obtain a loud or a soft note by striking a given key, but the pitch is the same.

With a swing or a pendulum we take it for granted that, by adjusting the length of the spring, we can make the frequency of oscillation anything we like. What the quantum theory says is that this is not really the case; only certain energies are possible. However, these quantum restrictions are too small to matter in a clock or a child's swing; for them to be noticeable we must deal with matter at the atomic level.

If we are dealing with an atom vibrating in a solid, or a vibrating oxygen molecule, we find that only certain amplitudes of vibration, and therefore only certain energies, are possible. Although we can ignore the quantization for visible objects like swings, when we get to the atomic scale the quantum effects must be taken into account. This explains why scientists failed to find any evidence for the quantum theory until the twentieth century. All their work had been done on much too large a scale.

The quantum theory had a rather obscure beginning, which explains why in spite of its great importance it was not accepted very quickly. It was put forward to explain some results on how the energy of the radiation emitted by a hot body varies with the wavelength or frequency of the radiation. Careful experiments on this were carried out around the turn of the century, particularly by a number of physicists in Berlin, and attempts were being made to explain the results. This was first done successfully by Max Planck, whose theoretical treatment of the radiation results was found to have much broader implications than he realized at first.

Max Karl Ernst Planck (1858–1947; Fig. 21) was born in Kiel, and studied at the Universities of Munich and Berlin. Here he came under the influence of Helmholtz, whose important work on thermodynamics we met in Chapter 2. Planck became professor of physics at Berlin in 1889, and much of his early research was in thermodynamics. On 25 October 1900 there was a scientific meeting in Berlin at which some new results on radiation were presented to the Academy of Sciences. At the meeting Planck suggested an empirical equation that seemed to fit the results better than any that had been suggested earlier. By the word empirical is meant that Planck designed the equation just so that it would fit the data. At this stage he had no theoretical reason at all for suggesting it.

It turned out that Planck's empirical equation fitted the data perfectly, and Planck proceeded to give it a theoretical justification. The essence of the treatment

Fig. 21 Max Planck (1858–1947), who was professor of physics at the University of Berlin from 1889. In 1900 he explained the distribution of energy in radiation in terms of the idea that energy comes in small packets, or quanta. For this work he was awarded the Nobel Prize in Physics for 1918.

he produced was that a solid body consists of an array of atoms each one of which is constantly vibrating, obeying the same laws that we have been discussing. He considered the distribution of the energy that these oscillators could have, and carried out a statistical treatment, making use of the methods that had been worked out by Boltzmann. As we saw in the last chapter, Boltzmann had often treated energy as coming in small packets, purely for mathematical convenience. At the end of his derivations he always required the energy packets to become of zero size.

Planck followed this procedure, and assumed that if the atoms were oscillating with frequency ν their energy came in packets of size $h\nu$, where h is a constant—it is now known as the *Planck constant.* In other words, the energy of the oscillation can be 0, $h\nu$, $2h\nu$, $3h\nu$, ..., but nothing in between. In his derivation Planck did not carry out the final step of making the energy packets go to zero, as Boltzmann had done, and he did not at first appreciate the great significance of this omission. His paper giving this treatment was presented to the German Physical Society on 14 December 1900, and this is often regarded as the birthday of the quantum theory.

In view of the fact that we now accept Planck's quantum theory as correct and of great importance we would have expected his 1900 analysis to have attracted immediate attention. We would at least have expected some discussion of it, and perhaps criticism. In fact, for about five years hardly any notice at all was taken of Planck's paper. The main reason for this is that the theory was presented only as applying to the radiation from solids, and its wider implications were not recognized until later. Most physicists at the time were working on what appeared to be much more exciting problems than radiation from hot bodies, such as radioactivity, X-rays, and the electron.

When attention was finally paid to Planck's theory, it was at first usually unfavourable. The criticisms that were made generally suggested not that Planck had formulated a quantum hypothesis that was unacceptable, but rather that he had just made a silly mistake in his mathematics.

An important contribution made by Planck in the same publication was to evaluate the constants h and k_B from the experimental data on the spectral distribution. It is surprising that Boltzmann had expressed his results in a way that did not involve what we now call the Boltzmann constant k_B, and so had never obtained a value for it; Planck was the first to do so. He also, from his theory and the data, obtained a value for the charge on the electron, which we now call the elementary unit of charge. The values he deduced are in reasonable agreement with the values accepted today. This contribution was of great importance, apart from the fact that it provided strong support to Planck's procedures.

At first Planck thought that his quantum theory applied only to oscillators—for example, to the atoms in a solid, which could possess energy only in multiples of $h\nu$. Radiation itself was not considered to be quantized until 1905, when Einstein made that important suggestion. In 1905 he published three papers of great importance, and the one that actually led to his 1921 Nobel Prize was on the quantization of radiation. His idea was that light of frequency ν can be regarded as if it were a beam of particles, having no mass but each one having energy $h\nu$.

Modern textbooks often say that Einstein's 1905 paper on quanta of radiation was based on Planck's 1900 paper on the quantization of oscillator energies. Only someone who had not looked at the paper could write that. In fact, Einstein made only a passing reference to Planck's work, which at the time he did not find convincing, and he made no use of Planck's constant h. Instead his proposal that radiation is quantized was based on a number of other considerations, one of which was what is called the *photoelectric effect*. This refers to the fact that if light of suitable frequency strikes a solid such as a metal, electrons are emitted.

This effect had been discovered in 1887 at the University of Kiel by Heinrich Rudolf Hertz (1857–1894) in the course of his famous experiments, inspired by Maxwell's electromagnetic theory of radiation, which eventually led to radio transmission. Rather paradoxically, it was a series of his experiments that convincingly confirmed Maxwell's theory of radiation, based on the idea that light involves vibrations, that also produced some of the first evidence for the particle nature of radiation. Two years later, in 1889, detailed studies of the emission of electrons under the influence of light were reported by Johann Elster (1854–1920) and Hans Geitel (1855–1923). These men were often known as the Castor and Pollux of science, or as the Heavenly Twins, since they were inseparable both in their private lives and in their teaching and research. They had been school friends, and both became teachers at a Gymnasium in Wolfenbüttel near Braunschweig (Brunswick). When Elster married and moved into a house Geitel joined the couple, and the two friends carried out research in a laboratory they established in the house. They were often confused with one another, and a man who somewhat resembled Geitel was once addressed by a stranger as 'Herr Elster'. His reply was 'You are wrong on two counts: first, I am not Elster but Geitel, and secondly I am not Geitel.'

The experiments done on the photoelectric effect revealed a rather remarkable result. If the frequency of the light was less than a certain value ν_0 there was no emission at all, however strong the intensity of the light. Moreover, if electrons were emitted, their energy was proportional to the difference between the frequency ν of the light and the critical frequency ν_0. It did not depend at all on the intensity of the light, which merely affected the number of electrons emitted and not their energy. Einstein realized that the results on the photoelectric effect are inexplicable in terms of the wave theory of radiation, according to which radiation of sufficiently high intensity, irrespective of frequency, would be able to eject electrons. With the particle theory, on the other hand, the explanation is straightforward (Fig. 22). A particle of radiation has energy $h\nu$, and is capable of ejecting an electron provided that $h\nu$ is sufficiently large; that is, that the frequency is large enough. At first these particles of radiation were just called quanta of radiation, which is still sometimes done, but in 1926 the American chemist Gilbert Newton Lewis (1875–1946) suggested the name *photon*, by which name they are now commonly called.

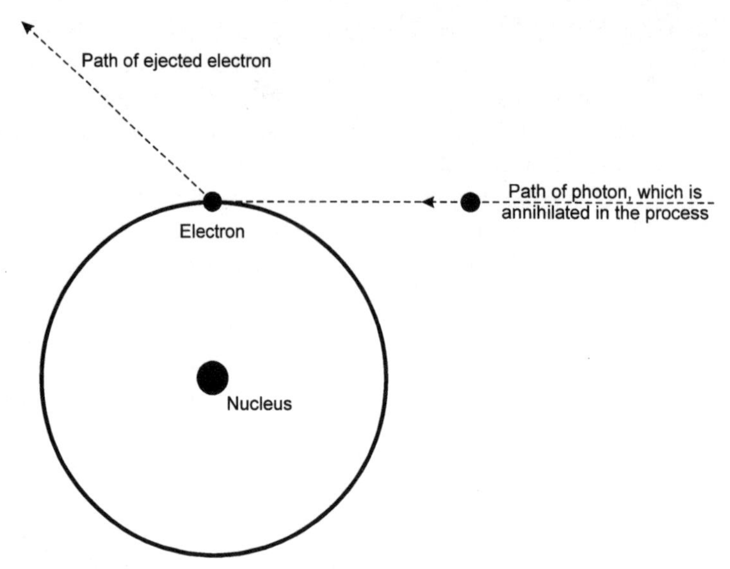

Fig. 22 Explanation of the photoelectric effect. An electron is shown in its motion around the nucleus, and a quantum of radiation (a photon) interacts with it. The energy of the photon is $h\nu$, where ν is the frequency of the radiation. A certain minimum energy, called the *work function*, is required to remove the electron. If the photon does not have this energy, nothing can happen; increasing the intensity of the light makes no difference, since it only increases the number of photons, and at a given frequency of light they all have the same energy. If $h\nu$ is greater than the work function, the photon will remove the electron from the atom and will be annihilated—think of the electron as just a bundle of energy. Any excess energy will be in the escaping electron.

Einstein's paper is often referred to today as his paper on the photoelectric effect, but much of the paper is concerned with other matters, such as the way in which atoms and molecules absorb and emit radiation. In other words, it was important in leading to explanations of the detailed nature of spectra.

Einstein's suggestion that light can act as a stream of particles seemed rather surprising at the time, and to appreciate this we should know a little about the development of the theory of light, which has had a complicated and troubled history. Over the years there have been two theories of the nature of light. One was that light is a stream of tiny particles, which used to be called corpuscles. The other was that it has wave motion, involving vibrations that are at right angles to the direction of propagation of the light. Corresponding to the wave there is a particular wavelength, denoted by the Greek letter λ (lambda), and a frequency, denoted by ν (nu). The two are related to the speed of light by the equation

$$\text{speed of light } (c) = \text{wavelength } (\lambda) \times \text{frequency } (\nu).$$

This is a relationship of fundamental importance which in an appropriately modified form applies to every kind of wave motion.

Newton gave careful consideration to both the corpuscular and wave theories, and did not completely discard either of them. Indeed, he sometimes favoured ideas that were partly wave and partly corpuscular, which is rather remarkable since that is the modern view. Newton's main difficulty with the wave theory is that he could not see how it could explain the fact that light travels in straight lines; he thought that when light encountered the edge of an obstruction the waves would be sent out in various directions. There is indeed some spreading of this kind (called diffraction), but the spreading is very slight and hard to detect. Newton changed his views from time to time, but on the whole he favoured the corpuscular theory. Because of his great prestige, this was the theory generally accepted until the beginning of the nineteenth century, when the situation changed.

Early in that century the British investigator Thomas Young (1773–1827) did important work on colour perception by the human eye and also on the theory of light. In careful experiments he was able to detect the spreading of light when it passes through narrow slits, finding that what are called *diffraction bands* or *interference bands* were formed. These results seem only to be explicable by the wave theory. In addition, Maxwell's theory of electromagnetic radiation was so convincing and successful that it seemed impossible that it could be wrong.

Einstein's suggestion in 1905 that light has particle properties thus seemed incredible, or at least surprising. Also, there appeared to be a logical difficulty with Einstein's theory: the energy of radiation $h\nu$ is expressed in terms of its frequency ν, but frequency only makes sense in terms of the wave theory.

In his 1905 paper Einstein anticipated these objections and answered them. He began his paper by admitting that the wave theory of light was well established and was not to be displaced as far as properties such as diffraction and interference are concerned. But, he continued, such properties, in which light interacts with matter in bulk, relate to *time averages*. It is quite possible that the wave theory will prove inadequate whenever there are *one-to-one effects* such as the photoelectric effect and the emission and absorption of radiation. In other words, what is needed is a *dual* theory of radiation. One way of explaining this is to say that light travels as a wave, but that when it arrives at its destination it behaves like a particle. For diffraction and interference the wave properties are relevant; for the photoelectric effect and for emission and absorption we must regard the radiation as behaving like particles. Einstein's suggestion that radiation can exhibit particle properties received strong support later, in 1923, when the American physicist Arthur Holly Compton (1892–1962) observed the scattering of X-rays.

At first Einstein was not convinced that oscillator energies are quantized, but only that radiation is quantized. Later he became convinced that Planck was quite right: oscillator energies must also be quantized. He became even more convinced of this when in 1907 he worked on the specific heats of solids, and found that he could give a good explanation of them by using Planck's treatment of vibrations in solids.

Even Planck had difficulty in accepting Einstein's theory of the quantization of radiation—or indeed in accepting that his own work really required us to think about things in quite a different way. To Planck the idea of quanta of energy was no more than an *ad hoc* hypothesis to explain black-body radiation; he was slow to realize that the quantum theory has a much wider significance. It was in his nature to be conservative, and having been brought up on classical physics he found it difficult to renounce the old ideas. Einstein deserves credit not only for introducing the idea of the quantization of radiation, but for persuading Planck that his theory was correct and important.

As time went on scientists came to realize that the quantum theory has many applications, and that it explained a variety of results that were impossible to explain without it. We have seen in Chapter 5 that specific heats showed rather inexplicable behaviour, in that they could not be explained by the kinetic theory that was used in the nineteenth century. Maxwell was one of the first to call attention to this in the case of the specific heats of gases, and as a result he had some doubts about the validity of the theory. The difficulties arose from the fact that energy was regarded as continuous, and when the quantization of energy was recognized the difficulties vanished.

With the specific heats of solids it had been found during the later years of the nineteenth century that they became small as the temperature was reduced, a fact that also seemed inexplicable. In 1907 Einstein showed, however, that if the quantization of the energy of vibration of the atoms of the solid is taken into account this can be explained, and he developed a detailed theory that fitted the data much better. Then in 1911 he worked out a similar theory that explained the specific heats of various types of gases.

There was a further interesting development in 1913 when Einstein and his research assistant Otto Stern (1888–1969) published a paper of great significance in which they suggested for the first time the existence of a residual energy that all oscillators have, even at absolute zero. They called this residual energy the *Nullpunktsenergie*, which is usually rather unsatisfactorily translated as the *zero-point energy*. They deduced from the specific-heat data for hydrogen gas that for an oscillator of frequency ν the zero-point energy would be $\frac{1}{2}h\nu$. The appearance of the $\frac{1}{2}$ in this expression appeared surprising at the time, but it has been confirmed by experiment and later by quantum-mechanical theory.

The next important event in the history of the quantum theory was its application to the structure and properties of atoms. This was first done successfully in 1913 by the Danish physicist Niels Bohr (1885–1962). Bohr was born in Copenhagen, and studied at the University of Copenhagen where he obtained his doctorate in 1911. He then worked for a short period at Cambridge with J. J. Thomson and then at the University of Manchester with Ernest Rutherford. It was there that he began the development of a theory of the hydrogen atom and of similar atoms. After his return to Copenhagen in 1913 he completed this work. In his theory he showed that if he took into account the fact that the energy of the electrons in their orbits was quantized, he could for the first time interpret the properties of the atoms, including the details of their spectra. This work was important in leading the way to the more precise theories based on the new science of quantum mechanics. Bohr was awarded the Nobel Prize in Physics in 1922.

Bohr's theory of the hydrogen atom provided a foundation for the understanding of the spectra of all atoms. In the nineteenth century it was impossible to explain spectra. Why should spectral lines appear in odd places in the spectrum? For example, a spectroscopist looking at the spectrum of a sodium flame observes a sharp line, called the D line, in the yellow region. Why is the light emitted particularly at this frequency and not at other ones? It is because only certain electronic orbits are allowed by the quantum theory. The sodium atom happens to have two electron orbits differing in energy by an amount that corresponds to this yellow line.

The Bohr theory of the atom was certainly a remarkable achievement, since it provided a general interpretation of atomic structures and spectra, which previously had seemed quite incomprehensible. However, even with the improvements that were made to the theory, serious difficulties remained, as Bohr himself freely admitted. For one thing, it was felt to be arbitrary simply to add quantum restrictions to the old mechanics. Also, it did not seem possible to extend the Bohr theory in such a way as to give a satisfactory treatment of atoms containing more than one electron, or even of the simplest of molecules. What was needed instead was a new mechanics in which the quantum restrictions would emerge as a mathematical necessity, and would explain the more complicated molecular systems.

This was done by several physicists in the 1920s, and the result was what we call quantum mechanics. The first to do so was Werner Heisenberg (1901–1976), who obtained his doctorate in 1923 at the University of Göttingen, under the direction of Max Born (1882–1970). He remained doing research with Born, and in the spring of 1925 he developed a treatment of mechanics which made use of an unusual type of mathematics, known as matrix theory. Born had contributed greatly to the work, but being a generous man suggested that Heisenberg should publish it in his name only. The theory, referred to as *quantum mechanics*, was soon recognized as being a great advance, and Heisenberg won the 1932 Nobel Prize in Physics for his system of quantum mechanics. It was always a matter of regret to him that Born had no share in this honour. The award of the prize jointly to Born and Heisenberg would indeed have been more appropriate. Not only had Born initiated the idea of a quantum mechanics: he had shown the true significance of Heisenberg's obscure mathematics. A less generous man would have delayed the submission of Heisenberg's paper until a joint paper of broader significance could be prepared. Born did receive a Nobel Prize much later, in 1954, for his work on the quantum mechanics of collision processes.

Another contribution of Heisenberg, made shortly after his matrix mechanics, has proved of great importance. This was his *uncertainty principle*, or principle of indeterminacy. According to it, the position of a particle and its momentum (mass times velocity) cannot both be measured precisely and simultaneously. We can express this differently by saying that the position and the energy of a particle cannot accurately be determined by experiment at any one time. Heisenberg arrived at the principle by carrying out an imaginary experiment in which a beam of light is used to determine the position and momentum of an electron. Because the light disturbs the electron, if we try to measure the momentum or energy we cannot do so without disturbing the electron, so that it moves and its position becomes unknown.

Shortly after the appearance of Heisenberg's first paper on quantum mechanics, an equally important contribution was made by Paul Adrien Maurice Dirac (1902–1984). Dirac was trained as an electrical engineer at the University of Bristol, and being unable to obtain employment secured in 1923 a research scholarship to work for his PhD degree at Cambridge. Late in 1925 he published an alternative formulation of the principles of quantum mechanics, a formulation that was equivalent to Heisenberg's mechanics but more comprehensible and ultimately more useful. Dirac's first paper on quantum mechanics was followed by many others of great importance. In 1926 he applied his methods to the hydrogen atom, and in papers that appeared in 1928 he combined his system of quantum mechanics with the theory of relativity. The resulting theory was an extremely powerful one which had far-reaching consequences. For example, Dirac was able to deduce that an electron can spin in just two different ways, a conclusion that had already been reached on the basis of experiment. Dirac also predicted from his theory the existence of an elementary particle having the mass of the electron but a positive charge. This particle, the *positron*, was discovered in 1932 by the American physicist Carl David Anderson (1905–1991).

Dirac wrote an important book, *The Principles of Quantum Mechanics*, which first appeared in 1930 and exerted a wide influence. In 1933 he was awarded the Nobel Prize for his work. When he heard that he had been selected for the Nobel Prize, he wanted to refuse it, saying that he disliked publicity. Rutherford told him that he would experience more publicity if he refused it, and that convinced him to accept. Dirac was a typical 'absent-minded professor' about whom many stories are told, and some of them may even be true. Once, when asked if he could play any musical instrument, he replied, after some thought, 'I don't know—I've never tried.' On another occasion while discussing physics with a colleague he idly watched a woman who was knitting. He finally told her that he had been working out the mathematics of knitting, and had discovered that there was an alternative procedure, which he explained. 'Yes,' she replied dryly, 'it is called purling, and has already been discovered.'

There were some other formulations of quantum mechanics, which in the end turned out to be equivalent to the theories of Heisenberg and Dirac, but which involved wave theory; as a result this formulation of quantum mechanics is called *wave mechanics.* We have seen that Einstein showed that radiation has both particle properties and wave properties. The converse idea was put forward in 1923 that particles such as electrons can also have wave properties. This suggestion came from the French physicist Louis Victor, Prince de Broglie (1892–1987). He had first intended to become a civil servant and at first studied history. Later, partly under the influence of his elder brother Maurice, Duc de Broglie (1875–1960), who was a distinguished experimental physicist, Louis studied physics and philosophy for his PhD degree at the Sorbonne.

De Broglie suggested the converse hypothesis, that particles such as electrons can have wave properties. By making use of equations for radiation, he suggested an equation that related the wavelength of the wave associated with a particle to the mass of the particle and its speed.

When de Broglie presented his PhD thesis in 1924 he had some trouble with his examiners. One of them dismissed the ideas as 'far-fetched'. However, de Broglie had shrewdly sent a copy of his thesis to Einstein, who returned an enthusiastic endorsement which included the phrase 'He has lifted a corner of the great veil'. In spite of this, the examiners remained sceptical of the work but somewhat reluctantly awarded the degree. Van de Graaff (1901–1969), the inventor of an electrostatic generator that bears his name, was present as a student at the defence of the thesis, and some time after the Second World War remarked—presumably having in mind Winston Churchill's famous remark about the RAF in the Battle of Britain—'Never has so much gone over the heads of so many.'

Other prominent physicists were also initially sceptical of de Broglie's theory, and he did not receive his Nobel Prize until 1929, by which time the theory had been confirmed experimentally. The first experiments to support the theory involved the *diffraction of electrons.* According to de Broglie's theory, electrons accelerated by a potential of about 100 volts should have wavelengths similar to the interatomic spacing in crystals, and are therefore suitable for diffraction experiments. In January 1927 Clinton Joseph Davisson (1881–1958) succeeded in demonstrating the diffraction of electrons by a single crystal of nickel. In May of the same year, at the University of Aberdeen, George Paget Thomson (1892–1975) and his research student Andrew Reid observed the diffraction of electrons by thin films. Ten years later Davisson and Thomson shared the Nobel Prize in Physics for this work. It has been commented that J. J. Thomson was awarded the 1906 Nobel Prize for showing that the electron is a particle, while his son G. P. Thomson won the 1937 Nobel Prize for showing that it is a wave. Of course, as we have seen, it is both, and both Nobel Prizes were well merited.

De Broglie's idea played no part in the quantum-mechanical treatments of Born, Heisenberg, and Dirac. It did, however, strongly influence the thinking of Erwin Schrödinger (1887–1961), who developed a wave mechanics that eventually turned out to be mathematically equivalent to the quantum-mechanical theories of Heisenberg and Dirac. His quantum mechanics has been particularly popular with chemists, since it provides a more easily visualized representation of atomic structure, in contrast to the rather formal approaches of Heisenberg and Dirac. Schrödinger, an Austrian by birth, became professor of physics at the University of Zurich in 1921, and it was there that he did his original work in wave mechanics.

Schrödinger's wave mechanics evolved directly from de Broglie's ideas, which he had first thought to be 'rubbish' until he was persuaded otherwise. In the words of the physicist Hermann Weyl (1885–1955), Schrödinger obtained his inspiration for wave mechanics while engaged in a 'late erotic outburst in his life'. His amorous exploits were somewhat remarkable, at any rate for a scientist. His unprepossessing appearance, with his thick spectacles, hardly corresponds to the popular idea of a Lothario or Casanova, but such he was. In 1920 he married Annemarie Berthel, and although their relationship was punctuated by many stormy episodes they remained together until his death. Schrödinger had no children by his wife, but he had at least three illegitimate daughters.

His wave mechanics was begun during one of his amorous adventures in late 1925, when he stayed at a holiday resort with a mistress while his wife remained in Zurich, and one wonders if the theory was the only conception at that time. The woman involved in that particular encounter has not been identified, but it would seem that she deserved at least a Nobel award as Best Supporting Actress.

In Schrödinger's wave mechanics an electron in an atom is treated as a wave rather than as a particle orbiting round the nucleus, which was Bohr's idea. He did not *derive* his wave equation—that is to say, he did not arrive at it by a formal mathematical proof. Instead he proceeded by analogy with the equations used in Maxwell's electromagnetic theory for wave motion in ordinary radiation. His wave equation is an equation involving what is called a *wavefunction*, and an energy E. Quantum restrictions are not introduced arbitrarily, but appear as a direct consequence of the wave equation, for which no mathematical solution is possible unless the energy has one of a number of permitted values. The restrictions as to the orbits in which an electron may move are a consequence of the need for a wave to fit into the right amount of space.

At first Heisenberg and Dirac were critical of Schrödinger's wave mechanics. Max Born approved of it from the start, and in 1926 provided a physical interpretation of the wavefunction. For a problem in atomic or molecular structure, the wavefunction according to Born is related to the probability that an electron is present in a particular small region of space. Under most situations the probability is simply represented by the square of the wavefunction.

The misgivings of Heisenberg and Dirac about Schrödinger's theory evaporated when later in 1926 Schrödinger proved that his and the other formulations were mathematically equivalent. Starting with Heisenberg's equations, which involve matrices representing physical properties, Schrödinger showed that each physical property can be replaced by an appropriate mathematical operator, and that his wave equation was then obtained. Modern quantum-mechanical calculations are often based on this procedure.

Heisenberg's uncertainty principle, and Born's interpretation of the square of the wavefunction as representing a probability, are important components of what came to be called the *Copenhagen interpretation of quantum mechanics*, since Niels Bohr, professor at Copenhagen, had much to do with formulating this point of view. The Copenhagen interpretation implies a lack of complete determinism, in that future events do not follow inevitably from past conditions, pure chance playing some role. Most physicists today accept this interpretation, but Einstein and Schrödinger took strong exception to it.

Einstein's objection is summarized in his often-quoted statement that 'God does not play dice.' In a number of forceful but always friendly arguments with Bohr, Einstein tried to devise ways of circumventing the uncertainty principle, but Bohr was always able to show that he was in error—sometimes by invoking Einstein's theory of relativity.

Schrödinger was an intensely emotional person, and this often led to rather bizarre personal behaviour. At the beginning of the Second World War he became director of the Institute for Advanced Studies in Dublin, but in 1945 he resigned, the reason being that he had become involved in a quarrel about the way his office was being cleaned. He usually dressed unconventionally, and as a result on several occasions had difficulty gaining admission to scientific meetings, and sometimes to lectures he was to give himself.

When the Copenhagen interpretation was put forward he did not just object to it, as a few other scientists did—he found it deeply distressing. With regard to Born's probability explanation, Schrödinger would say: 'I can't believe that an electron hops about like a flea.' In one exchange with Bohr he said that since people were giving these interpretations to his wave mechanics, 'I regret having been involved in this thing.' To Schrödinger an electron has wave properties and is not to be regarded as a particle darting about. In his view electronic orbitals are to be considered in terms of wave properties, not as the movement of particles.

We saw in previous chapters that there was much controversy about the role of chance with reference to the second law of thermodynamics. In the end, chance has been accepted as giving the correct explanation. We have now seen again that in the new quantum mechanics chance is all important. We cannot say exactly where an electron is, but only that it has a certain chance of being in a particular place.

Chance thus again plays a fundamental role. It is the essential ingredient of the structure of matter, and of how processes occur. In fact, it plays a dominant role in everything, including ourselves and our behaviour.

EIGHT

Energy equals mc^2

Anyone who expects a source of power from the transformation of atoms is talking moonshine.

Lord Rutherford, said to James Chadwick in 1933
(within a year or so he had changed his mind)

Besides proposing that radiation is quantized, Albert Einstein (Fig. 23) also formulated in 1905 his special theory of relativity. Since this theory, and the general theory of relativity which he developed in 1916, are the investigations for which he is most famous, it is at first sight surprising that the 1921 Nobel Prize was awarded to him for his work on the quantum theory. The reason is that at the time of the award there was still not considered to be sufficient evidence for relativity theory. Indeed, the Nobel committee had so many doubts about Einstein's work that the decision to give him the 1921 award was not made until

Fig. 23 Albert Einstein (1879–1955), one of the greatest scientists of all time. He is best known for his theory of relativity, but his work on the quantum theory (Chapter 7) was also of great importance. His 1921 Nobel Prize was specifically for his work on the quantization of radiation.

1922. Another feature of Einstein's Nobel Prize is unusual. For some years he and his first wife Mileva were not getting along well, but she was reluctant to leave him. When it became fairly certain that Einstein would win the Nobel Prize he made a bargain with his wife, which they kept, that he would give her the prize money as compensation for her leaving him.

Albert Einstein (1879–1955) was born in Ulm, Germany. His father was a not very successful electrical engineer who had constant business troubles that caused the family to move frequently. After an undistinguished school career Albert entered the Swiss Federal Institute of Technology in Zurich at the age of 17. He was not a particularly good student, and partly because he caused personal offence to some of his teachers he was unable to obtain an academic appointment. As a result he had to settle for a junior and rather dull position in the patent office in Berne. In 1901, disliking the political atmosphere in Germany, he renounced his German citizenship in favour of Swiss citizenship. In 1903 he married Mileva Maric, who had been a fellow student of physics at the institute. In the previous year their daughter Lieserl had been born; they put her out for adoption and lost touch with her so that her fate is unknown. The Einsteins later had two sons.

Einstein did competent work in the patent office, and since the position was by no means demanding he had the leisure to carry out a good deal of research and to publish a number of scientific papers. The year 1905 was an *annus mirabilis* for him, since in that year he published six papers, two of which were worthy of Nobel Prizes. The first of these was on the quantization of energy, which we considered in the last chapter and which much later did actually get him the Nobel Prize. The second was on the Brownian movement, and the third on the special theory of relativity. All three papers appeared in the *Annalen der Physik*, an important German scientific journal, in late 1905.

The special theory, and the general theory of relativity which Einstein published in 1916, are regarded as among the greatest of all contributions to science. They have had a profound influence on the way scientists think about fundamental problems. In addition, some aspects of the theories have had remarkable practical consequences, leading in particular to an important way of producing energy by means of nuclear processes.

The special theory of relativity is based on formulating the laws of physics in a way that is common to all observers under any conditions. To understand the theory we need to know something of its background. When the wave theory of light was being developed it was considered by most investigators that there must be some medium, pervading all space, which carried the light waves. Ripples on the surface of a pond are transmitted by the water, and it seemed that it was necessary for the light waves to exist in some medium. This was called the *ether* (sometimes, especially formerly, spelt 'aether'), and it was often called the luminiferous (i.e. light-bearing) ether. Maxwell's electromagnetic theory strengthened the belief in the ether, since light, electricity, and magnetism are all associated with an electromagnetic field, and it seemed that the field must be sustained by some medium such as the ether. In his article on 'Ether' in the 9th edition of the *Encyclopaedia Britannica* (1875) Maxwell expressed very clearly his belief in the existence of an ether. He considered, however, that his theory of electromagnetic radiation was valid whether or not the ether exists, or whatever its nature is.

Astronomical observations about the motion of the stars and planets led scientists to the conclusion that these bodies must move through the ether without producing any disturbance. If this is the case it should be possible to determine the speed with which the Earth moves through the ether by making observations on the speed of light. Suppose that the Earth is travelling through the ether in a particular direction with a speed of v relative to the ether. If light, moving with speed c, travels in the same direction, its speed relative to the ether should, according to classical theory, be $v + c$. If, however, the light travels in the opposite direction, its speed relative to the ether should be $c - v$. The situation is analogous to a person swimming with a speed c in a river flowing at the rate v. If the swimmer travels in the same direction as the stream, the speed relative to the bank is $c + v$; if the swimmer travels in the opposite direction the speed is $c - v$. The relationships are easily worked out if the swimming is perpendicular to the flow, or if it is diagonal. In the case of measurements made of the speed of light in directions at right angles to each other the theory leads to the result that the ratio of the measured rates should be $\sqrt{(1 - v^2/c^2)}$.

Information about the ether should therefore be obtainable from measurements of the speed of light in two directions at right angles to one another. An experiment could be carried out in which a beam of light is sent out from a position in such a way that it strikes a mirror some distance away and is reflected back. The light from the same source could also be made to travel in a direction at right angles to the previous direction, and reflected by another mirror. By comparing the times of arrival of the two reflected beams it would then be possible to calculate v, the speed with which the Earth is travelling through the ether. An experiment of this kind had been suggested by Maxwell when developing his electromagnetic theory.

Such experiments were first carried out in 1891 by the American physicist Albert Abraham Michelson (1852–1931). Born in Strelno, which is now in Poland, Michelson was brought by his parents to the United States at the age of 4. He was educated at the Naval Academy at Annapolis, and following a tour of duty at sea was appointed an instructor there. Throughout his career he devoted much effort to improving the accuracy of the measurement of the speed of light. In 1880 he took a study leave and worked in the laboratories of the eminent physicist Helmholtz at the University of Berlin. In Helmholtz's laboratory he built an instrument called an interferometer that allowed two narrow beams of light to be emitted in directions at right angles to one another, and for the speeds to be compared in the two directions. He found that the speeds were exactly the same.

Michelson left the Navy in 1881 and became professor of physics at the Western Reserve University in Cleveland, Ohio. The professor of chemistry there was Edward Williams Morley (1838–1923), a man who, like Michelson, had a great enthusiasm for making accurate physical measurements; one of his achievements was to make an extremely accurate measurement of the relative mass of the oxygen molecule. The two men joined forces and together made many further measurements on the 'ether-drift' problem, always finding a null result. They tried with various orientations of the beams, but always found that the speed of light was independent of the direction of propagation. Their joint paper published in 1887 created a deep impression on physicists, and in the following year Michelson was awarded a national prize, the citation saying that it was 'not only for what he has established, but for what he has unsettled.'

This conclusion that the rates are the same is now generally accepted, and as we shall see is consistent with Einstein's theory of relativity. It is of interest, however, that the result did not go unchallenged. Morley later carried out some further experiments with the American physicist Dayton Clarence Miller (1866–1941) who was professor at the Case School of Applied Sciences in Cleveland. Some of their results seemed to yield a positive result, suggesting the existence of an ether, but Morley discounted them as due to experimental error. Miller, on the other hand, took them seriously and from 1902 to 1932 carried out many further experiments on his own, some of them at Mount Wilson Observatory in California. His announcement in 1925 of positive results attracted much attention, since they were regarded by some as a refutation of Einstein's theory of relativity. On the basis of this work Miller was awarded a prize by the American Association for the Advancement of Science. A reappraisal of his work in the 1950s suggested that his results were erroneous because of slight variations in the temperature of his equipment.

The null result obtained by Michelson and Morley was generally accepted, and various suggestions were made to explain it. Similar ideas were put forward by G. F. Fitzgerald and H. A. Lorentz. The Irish physicist George Francis Fitzgerald (1851–1901) was educated at Trinity College, Dublin, remaining there as professor for the rest of his short life. To explain the null result from the Michelson–Morley experiment he suggested that a body moving through the ether contracts slightly in its direction of motion, in proportion to its velocity. Such contraction would not be experimentally observable, since any measuring instrument would contract correspondingly and the length would appear to be unchanged. The idea was that the apparatus used in the Michelson–Morley experiment would change its dimensions in such a way as to compensate exactly for the expected change in the observed velocity of light.

A similar explanation was put forward by the eminent Dutch physicist Hendrik Antoon Lorentz (1853–1928), who for many years was professor of theoretical physics at the University of Leiden. He too suggested that the null results obtained by Michelson and Morley were due to the contraction of matter as it moves relative to the ether, and this contraction is now commonly known as the *Lorentz–Fitzgerald contraction*. In 1904 Lorentz developed a mathematical treatment of the contraction, concluding that if, for example, an electron has radius r_0 when at rest, its radius r when it is moving at speed v is given by

$$r = r_o \sqrt{(1 - v^2/c^2)}.$$

Since v is bound to be less than c the ratio v^2/c^2 is a fraction and therefore $\sqrt{(1 - v^2/c^2)}$ must be less than one; r is therefore less than r_o, corresponding to a contraction.

Lorentz then took this argument a stage further. He suggested that according to electromagnetic theory, the mass of an electron is inversely proportional to its radius. If, therefore, we represent the mass of the electron when it is stationary as m_o, its mass when moving with a speed v is given by

$$m = \frac{m_o}{\sqrt{(1 - v^2/c^2)}}.$$

We shall see that the same relationship was later given by Einstein in his special theory of relativity, so that Lorentz's work was of great importance in paving the way for relativity theory.

Einstein's great contribution in his special theory of relativity was to arrive at these and other relationships in a more satisfactory and comprehensive way. Einstein discarded the ether concept as unnecessary, and made two important postulates. One was that the laws of motion are exactly the same when they are determined by observers moving at different speeds. The second is that the velocity of light has a constant value, irrespective of how the source of the light is moving relative to the observer.

The ideas of the special theory are best understood by means of an analogy. Suppose that an observer is drifting along a river which is flowing at speed v, and that a signal light on the bank flashes a beam of light. To an observer on the bank the speed of the light would be c. An observer on a boat moving towards the signal, however, would according to classical physics measure the speed of the light relative to the boat to be $c + v$. If the observer was travelling away from the light the speed of light would appear to be $c - v$. According to Einstein's theory, however, the observer on the boat will always find the speed of light to have the value c, irrespective of the speed of motion v. This can be visualized in terms of the Lorentz–Fitzgerald contraction, the instruments used for measuring the relative speed of the light undergoing a change. As a result of the change the observer always finds the speed of the light to be the same, irrespective of the relative motion.

The relativity corrections involve the term $\sqrt{(1 - v^2/c^2)}$. These corrections are unimportant if v is less than a tenth of the speed of light. If v is one-tenth of the speed of light, that is if v/c is one-tenth, then $\sqrt{(1 - v^2/c^2)}$ is 0.995 so that the correction is only 0.5 per cent. On the other hand if the speed v is 99 per cent of the speed of light, the factor works out to be 1/7.1, but if the speed v is 99.9 per cent of the speed of light the factor is 1/22.4.

Einstein's theory led to the same formula as given by Lorentz to express the relationship between the mass m of a body moving with relative speed v and the rest mass m_o:

$$m = \frac{m_0}{\sqrt{(1 - v^2/c^2)}}.$$

If v is one-tenth of the velocity of light, that is v/c is one-tenth, then $\sqrt{(1 - v^2/c^2)}$ is 0.995 and the actual mass is $m_0/0.995 = 1.005m_0$. In other words, the effective mass is only 0.5 per cent greater than the rest mass. On the other hand, if the speed v is 99 per cent the speed of light, the effective mass is 7.1 times the rest mass. If the speed is 99.9 per cent of the speed of light the mass is 22.4 times the rest mass. An important point about these relativity corrections is that Lorentz had deduced them only for an electron; Einstein's derivations were more general, applying not only to an electron but to any particle.

Quite a few years before Einstein had presented his special theory there was some experimental evidence for the increase in mass of an electron when it moves at high speed. In 1900 the German physicist W. Kaufmann described a method for determining the ratio of the charge e to the mass m of an electron emitted as a beta particle (β-particle) from a radioactive source. These electrons are moving at high speeds, and Kaufmann obtained good evidence for a decrease in e/m as the speed of the electrons approached the speed of light. When the speed of the electron was less than about one-tenth of the speed of light e/m was found to be constant, but, in accordance with the theory, at higher speeds the decrease could be detected. Now that it has become possible to accelerate electrons to speeds that are not too far short of the speed of light, this increase in mass at high speeds has often been confirmed experimentally.

From this relationship between the effective mass and the rest mass, Einstein was easily able to show that the change in the mass, denoted by Δm, resulting from the change in speed is the kinetic energy divided by the square of the speed of light: $\Delta m = E_k/c^2$. Then, by a simple extension of this relationship, he proceeded, as he put it, to prove that 'the mass of a body is a measurement of its energy content'. In other words, when the energy of a body is changed by an amount E, no matter what form the energy takes, the mass of the body will change by E/c^2. It is thus possible to write the equation

$$E = mc^2.$$

This means that under certain circumstances, energy and mass can be interconverted. For example, a process may occur in which there is a loss of mass, in which case the gain of energy is given by this equation. Alternatively, when we accelerate a body such as an electron we are providing it with energy; some of it goes into its increased kinetic energy, but some of it is converted into its mass.

For ordinary chemical reactions this mass–energy relationship can be ignored, the mass changes being much too small to measure. For example, if 2 grams of hydrogen react with 16 grams of oxygen to form water (the chemical equation is $H_2 + \frac{1}{2}O_2 \rightarrow H_2O$), the heat (i.e. energy) evolved is found experimentally to be 241 750 joules. From the $E = mc^2$ equation it can be calculated that there must be a decrease in mass of 2.7×10^{-9} grams. Even if it were a thousand times as much as this it would be very difficult to detect.

When nuclear reactions occur, on the other hand, there can be an enormous production of energy. Since nuclear reactions occur in the Sun and in all the stars, they are important for understanding the ultimate sources of energy. In particular, our deductions about how the universe must have been created require us to take into account the mass–energy transformations, as expressed by the formula.

Surprisingly, the idea of the conversion of mass into energy was considered quite early in the twentieth century, even before Einstein's 1905 paper on the special theory appeared. The German physicists Johann Elster and Hans Geitel, whom we met in the last chapter in connection with the photoelectric effect, first called attention to the possibility. After the discovery of radioactivity by Henri Becquerel in 1896 they became interested in the source of the energy of the radiation that was emitted by a radioactive substance. They carried out experiments to test various possibilities, and concluded in 1903 that the energy cannot come from any outside source, but must come from the interiors of the atoms themselves. Because of this conclusion, the house in Wolfenbüttel where Elster and Geitel lived (along with Elster's wife) and did their work bears a memorial plaque on which the two men are honoured as the *Entdecker der Atomenergie*, the discoverers of atomic energy.

To get some idea of the enormous amounts of energy released when mass is converted into energy, suppose that you were skiing down a hill and developed a high speed. Suppose, hypothetically, that you attained the speed of light. It follows from Einstein's formula that if you hit a tree, the liberation of energy due to your complete annihilation would be equivalent to that produced by about 40 nuclear warheads. There is another way of appreciating the enormous amounts of energy that are tied up in nuclei. If all the energy of 1 gram of matter could be released, it would be enough to drive one of the largest liners across the Atlantic.

Examples of how mass can be converted into energy are provided by radioactive disintegrations, and we will have to digress a little to consider atomic nuclei and how they are composed. At first it was assumed that atoms are hard balls of matter, but work in the laboratories of Ernest Rutherford (1871–1937) led to the conclusion that an atom has a nucleus which is much smaller than the atom itself. Associated with the nucleus are electrons, which give the atom its size. Typically, a nucleus occupies only about one-million-millionth (10^{-12}) of the volume of the atom. To gain some idea of the small size of the nucleus of an atom compared with the size of the atom itself, imagine an atom magnified to a radius of 10 metres—roughly the size of a large bus. The radius of the nucleus would be less than a millimetre—not much more than the dot at the end of this sentence. If a nucleus were magnified so that its radius is about the width of this page, the electrons would be more than a kilometre away.

It is the atomic nuclei that are involved in radioactive processes. It is now known that nuclei are composed of protons and neutrons, in various proportions.

A proton has a single positive charge, while a neutron has a zero charge. We can think of a neutron as a proton neutralized by an electron, which has an equal negative charge. At first it was thought that atomic nuclei were composed of protons and a certain number of electrons, but it is now known that the neutron, like the proton, is really an elementary particle. It was discovered and properly identified in 1932 by the British physicist James Chadwick (1891–1974). Although neutrons can now be detected reasonably easily, they escaped detection for some years. Since they bear no electric charge, neutrons can pass right through many types of matter without doing much to it and therefore without creating much of a disturbance—just as an invisible man whose mass was concentrated in a ball one-million-millionth of normal size would escape notice in a crowd provided that he did not move around too much. Neutrons can make themselves felt if they are caused to move at high speed by the use of a powerful accelerator such as a cyclotron. With modern techniques physicists can generate neutrons and experiment with them, so that we know a good deal about them. We even know that there are certain highly condensed stars, the pulsars, that consist almost entirely of neutrons. An interesting property of neutrons is that by themselves they are unstable, but that when combined with protons in an atomic nucleus they can be stable.

The simplest atom is the hydrogen atom, in which the nucleus has a single positive charge. Associated with it is an electron, so that the atom as a whole has no charge. In most of the hydrogen atoms on Earth the nucleus consists of just a single proton, with no neutron, but there are two other forms of hydrogen (Fig. 24). One of them (Fig. 24(b)) differs from ordinary hydrogen H in that the nucleus is about twice as heavy. The reason for this is that the nucleus consists of a proton and a neutron. This particular form of the hydrogen atom we call hydrogen-2 or *deuterium* and give it the symbol D; we can also write it as $^{2}_{1}H$, the superscript 2 meaning that there are two particles in the nucleus, and the subscript unity meaning that one of them is a proton, the other being a neutron. The well-

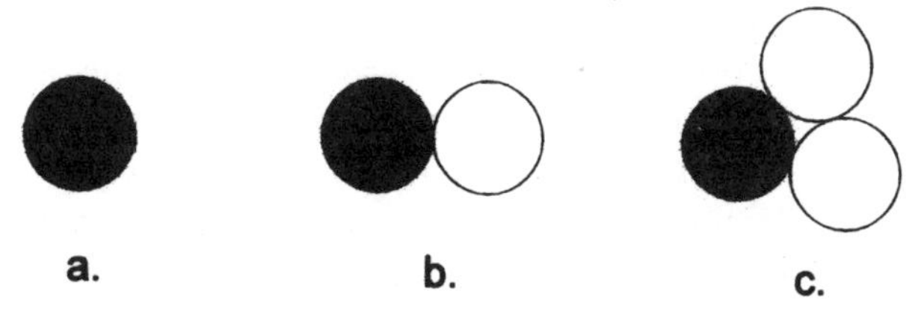

Fig. 24 The nuclei of the three isotopes of hydrogen, showing the numbers of protons (solid circles) and neutrons (open circles). (a) Ordinary hydrogen; the nucleus is a proton, and there is an orbital electron. (b) Deuterium, where the nucleus is a proton and a neutron. (c) Tritium, where the nucleus is a proton and two neutrons. In the neutral atoms, each of these nuclei has a single electron associated with it. In the old theory this electron is regarded as being in an orbit, but according to quantum mechanics it exists as a kind of cloud, referred to as an *orbital*.

known substance *heavy water*, used in some nuclear reactors, is made up of an oxygen atom combined with two deuterium atoms, and is written as D_2O.

There is yet another possibility for the hydrogen atom. A proton can become attached to two neutrons (Fig. 24(c)), so that its mass is three times as great as that of ^1_1H. We call this form hydrogen-3 or *tritium* and give it the symbol T or ^3_1H. This form of hydrogen is very unstable, and does not survive very long. The word *isotope* is used to refer to atoms whose nuclei have the same number of protons but which differ in the number of neutrons. Different isotopes of an element are always very similar in chemical and physical properties, which depend much more on the charge on the nucleus than on the mass.

Certain combinations of protons and neutrons, however, cause the nucleus to be unstable and to break up. For example, there is no nucleus composed of just two protons. There is one, however, that contains two protons and one neutron, and another that contains two protons and two neutrons. To make a neutral atom out of each of these, two electrons must be present. We now have a different chemical element altogether, since the chemical behaviour of an atom is determined by the number of electrons associated with the nucleus, which is equal to the number of protons in the nucleus. This atom which has two protons in its nucleus is called *helium*. The two forms of helium, which we write as ^3_2He and ^4_2He, are isotopes of helium (remember that the subscript tells us the number of protons, the superscript the total number of particles in the nucleus). Since the proton and the neutron weigh much the same, ^4_2He is about four times as heavy as an ordinary hydrogen atom, while ^3_2He is three times as heavy. Helium found in nature is almost all ^4_2He, the proportion of it being 99.999 86 per cent, so that the amount of ^3_2He is exceedingly small.

The element carbon, which plays an important part in living systems, exists in three known isotopic forms. The most common, with an abundance of 98.9 per cent, is $^{12}_6C$ or carbon-12, its nucleus consisting of six protons and six neutrons, so that the mass number is 12. There is also $^{13}_6C$, carbon-13, in which there is an extra neutron; its proportion is 1.1 per cent. A third form, $^{14}_6C$, has also been detected in tiny amounts.

For one of the lighter elements a useful rule of thumb is that the isotope in which the numbers of protons and neutrons are the same is the most common. This is true for oxygen, for which the most common isotope is $^{16}_8O$, with eight protons and eight neutrons. It is also true for nitrogen, where the common form is $^{14}_7N$, with seven protons and seven neutrons. When we get to the heavier elements, however, this is no longer the case. The only isotopes that are at all stable tend to have many more neutrons than protons. Here are some examples:

Radium	$^{226}_{88}Ra$	88 protons and 138 neutrons; 226 particles
Thorium	$^{230}_{90}Th$	90 protons and 140 neutrons; 230 particles
Uranium	$^{238}_{92}U$	92 protons and 146 neutrons; 238 particles.

Although these isotopes are the most stable of their respective elements, they do not live for ever, unlike other elements such as carbon and oxygen. The three elements radium, thorium, and uranium all show *radioactivity*, a phenomenon that was discovered in 1896 by the French physicist Antoine Henri Becquerel (1825–1908). His work was done with uranium, the most abundant radioactive element in nature. It is found to emit the nuclei of ${}^{4}_{2}$He atoms, which are known as alpha particles (α-particles); the process is called radioactive *disintegration*. To see what has happened we can do some arithmetic as follows:

$$ {}^{238}_{92}\text{U (92 protons + 146 neutrons)} \rightarrow {}^{4}_{2}\text{He (2 protons + 2 neutrons) + ?} $$

The product must contain $92 - 2 = 90$ protons, which means that it is the element thorium (remember that the name of the element is determined by the number of protons in the nucleus). It has 144 neutrons, the total number of particles being 234:

$$ \begin{aligned} {}^{238}_{92}\text{U (92 protons + 146 neutrons)} \rightarrow & {}^{4}_{2}\text{He (2 protons + 2 neutrons)} \\ & + {}^{234}_{90}\text{Th (90 protons + 144 neutrons).} \end{aligned} $$

The most stable form of thorium is ${}^{230}_{90}$Th and the isotope produced in this process, ${}^{234}_{90}$Th, is less stable since it has four neutrons too many. It also undergoes radioactive disintegration, but not in the same way as the uranium. Instead of giving off an α-particle, which is a helium nucleus, it emits a β-particle, which is an electron. Emission of an electron is equivalent to converting a neutron into a proton:

$$ \text{neutron} \rightarrow \text{proton + electron.} $$

When a ${}^{234}_{90}$Th (90 protons + 144 neutrons) nucleus emits a β-particle, the product nucleus therefore has 91 protons and 143 neutrons, the total number remaining at 234. We can thus write the process in terms of the balanced equation

$$ \begin{aligned} {}^{234}_{90}\text{Th (90 protons + 144 neutrons)} \rightarrow & \text{ electron} \\ & + {}^{234}_{91}\text{Pa (91 protons + 143 neutrons).} \end{aligned} $$

The element that has a nuclear charge of 91 is called protoactinium, and given the symbol Pa. In order to make the numbers in this equation balance we write the electron, which is a β-particle, as ${}^{0}_{-1}\beta$, the subscript -1 meaning that its positive charge is -1; the superscript 0 means that the tiny mass is effectively zero. Thus we write

$$ {}^{234}_{90}\text{Th} \rightarrow {}^{0}_{-1}\beta + {}^{234}_{91}\text{Pa.} $$

The two radioactive processes that we have considered are typical of many other spontaneous nuclear processes that occur. They differ in that in one of them a helium nucleus (α-particle) is emitted, in the other an electron. They also differ considerably in another way, namely the rates with which the disintegrations occur.

We are now ready, after this explanation, to consider the energies released when some of these processes occur. Ordinary radium undergoes the following process, the products being the elements helium He and the gas radon Rn:

$$ {}^{226}_{88}\text{Ra} \rightarrow {}^{4}_{2}\text{He} + {}^{222}_{86}\text{Rn.} $$

The following are the masses of the nuclei involved:

^{226}Ra	226.0312 u	^{4}He	4.0026 u
		^{222}Rn	222.0233 u
		sum	226.0259 u.

The unit, u, used here is the *atomic mass unit*; it is defined in such a way that the mass of the isotope ^{12}C (carbon-12) of the carbon atom is exactly 12 u. We see that the mass has decreased by $226.0312 - 226.0259 = 0.0053$ u. It is easy to calculate from Einstein's formula that one atomic mass unit corresponds to an energy of 931.5 million electron volts, MeV, and it follows that a decrease in mass of 0.0053 u will lead to the production of 4.9 MeV of energy. The *electron volt*, eV, is a convenient unit to use in work of this kind. It is the energy acquired by an electron when it passes through a voltage drop of 1 volt; 1 electron volt is roughly the energy that is carried by one photon of visible light. We get some idea of the vast amount of energy evolved in these nuclear transformations when we note that in ordinary chemical reactions, even in conventional explosions, the energy produced is only a few electron volts. With nuclear processes we are talking about millions of electron volts.

Two other types of nuclear transformation are of great importance. The first is called *nuclear fission*, and involves the breakdown of a nucleus into smaller nuclei. The first nuclear fission reaction to be discovered was brought about by introducing a neutron into uranium-235. The neutron was captured, with the formation of uranium-236, which breaks down into two nuclei, barium-145 and krypton-88. Since $145 + 88 = 233$ is three less than 236, there are three neutrons left over. Since the neutron has unit mass but no charge we write it as $^{1}_{0}$n, and the process is

$$^{235}_{92}\text{U} + {}^{1}_{0}\text{n} \rightarrow {}^{236}_{92}\text{U} \rightarrow {}^{145}_{56}\text{Ba} + {}^{88}_{36}\text{Kr} + 3\,{}^{1}_{0}\text{n} + \text{energy.}$$

The amount of energy released in this process is large, about 200 MeV. What is particularly noteworthy about this reaction is that a single neutron has produced a large amount of energy without any loss of neutrons from the system; in fact, three neutrons have appeared in its place. These neutrons will cause the breakdown of neighbouring uranium nuclei, with the production of more neutrons. When this is the situation we have what is called a *chain reaction*, and within a fraction of a second many of the uranium nuclei present will have undergone fission. A small number of neutrons introduced into uranium-235 will therefore lead to the production of vast amounts of energy in a very short time. In practice, the amount of energy released is never as much as it would be if all of the nuclei underwent fission, only a few per cent of the energy that is latent in the uranium-235 being released. Nevertheless, the energy produced is vastly more than in an ordinary chemical reaction, perhaps a million times as much.

This is the reaction that first made the commercial production of nuclear energy possible, and it has an interesting history. In December 1938, less than a year before the Second World War began, Otto Hahn (1879–1968) and Fritz

Strassmann (1902–1980), working at the Kaiser Wilhelm Institute in Berlin, observed that when uranium, the heaviest element then known, was bombarded with neutrons, one of the products was barium, a relatively light element. At once Lise Meitner (1878–1968) and her nephew Otto Robert Frisch (1904–1979) interpreted Hahn and Strassmann's observation as indicating that the neutrons were breaking the uranium nucleus into two nuclei of roughly equal mass. At the time both Meitner and Frisch were refugees from Germany; Frisch was working in Niels Bohr's laboratory in Copenhagen, and Meitner was working in Stockholm. They also calculated that during the process there was a substantial loss of mass and therefore, by Einstein's relationship, an enormous liberation of energy—much greater than in any nuclear process that had been previously observed.

When Frisch told Bohr about their conclusion Bohr struck his forehead and exclaimed 'Oh, what fools we have been. We ought to have seen that before.' Frisch did not at first know what to call the type of process that was occurring, and asked a biologist what term was used to describe the splitting of a biological cell into two daughter cells. 'Fission' was the answer, and in the Meitner–Frisch publication in the scientific journal *Nature* the word 'fission' appeared—not yet 'nuclear fission', which came later.

In Paris in the same year, Frédéric Joliot-Curie (1900–1958) and his wife Irène Joliot-Curie (1897–1956)—the daughter of Marie Curie (1867–1934), famous for her early work on radioactivity—found that there were more neutrons released in the process than had been used in the initial bombardment of the uranium. These secondary neutrons could disintegrate more uranium atoms, and thus set off a self-perpetuating chain reaction which would spread through the whole of the uranium present.

There is another kind of nuclear process, called *fusion*, which in principle leads to the production of even greater amounts of energy. In a fusion reaction the nuclei simply combine together, and an important example involves the hydrogen isotope deuterium. The mass of the nucleus of a deuteron ($^2_1$H) is 2.0141 u, while that of a helium atom ($^4_2$He) is 4.0026 u. If therefore two deuterons are forced together to form a helium nucleus, there is a mass loss of $4.0282 - 4.0026 = 0.0256$ u, and a corresponding release of energy of about 23.7 MeV:

$$^2_1\text{H} + {}^2_1\text{H} \rightarrow {}^4_2\text{He} + 23.7\ \text{MeV}.$$

This reaction took place extensively in the Big Bang, at the creation of the universe, and subsequently has occurred in the stars; this will be discussed further in Chapter 9. Its occurrence accounts for the fact that, of all of the matter present in the universe, one-quarter by weight is composed of $^4_2$He atoms (about 75 per cent is hydrogen, with only traces of all the other elements).

When attempts are made to make this reaction occur on Earth, a difficulty arises from the fact that the highest temperatures that can be attained are much lower than those in the Big Bang or in the stars, and are too low for $^4_2$He to be formed readily. Instead, the following two reactions, which produce less energy, occur preferentially:

$$^{2}_{1}\text{H} + ^{2}_{1}\text{H} \rightarrow ^{3}_{2}\text{He} + ^{1}_{0}\text{n} + 3.28 \text{ MeV}$$
$$^{2}_{1}\text{H} + ^{2}_{1}\text{H} \rightarrow ^{3}_{1}\text{H} + ^{1}_{1}\text{H} + 4.04 \text{ MeV}.$$

Another difficulty with the reactions is that the two positively charged nuclei repel one another and resist being brought together. Special techniques are therefore necessary, particularly to produce extremely high temperatures; it is because of these high temperatures that the processes are often called *thermonuclear*. In the stars the temperatures are sufficiently high that the fusion reaction to form $^{4}_{2}\text{He}$ is going on all the time. The enormous amounts of energy released in the fusion reactions keep the stars hot for billions of years, but eventually they will die when their nuclear fuel is used up.

Einstein's special theory of relativity was published in 1905, and his general theory in 1916. It was not for a few more years that his ideas became generally accepted. The energies generated in nuclear processes provided powerful support for the special theory but in 1919, before such data became available, some strong astronomical evidence had been obtained for the general theory. We need not go into that theory in this book; for our purposes it is enough to say that according to the general theory space becomes distorted in the neighbourhood of matter, the curvature of space accounting for gravitational attraction. Because of the curvature a beam of light is bent as it passes close to a solid body. The experimental verification of this bending was therefore important in leading to support for the theory.

The scientist who made the greatest contribution to testing the general theory of relativity was the British astrophysicist Sir Arthur Eddington (1882–1944). The theory made a prediction that stars that appeared to be near the Sun during a solar eclipse (but are actually, of course, much further away from us) would be displaced from the positions that had been observed for them at other times. This is because of the curvature of space near the Sun, as predicted by Einstein's theory. Until 1919 it had not been possible to confirm this prediction, but on 29 May of that year there was an unusually auspicious eclipse, one that might not have occurred again for hundreds of years. On 29 May of every year the Sun is in front of an unusually dense group of bright stars, known as the Hyades cluster. Without a solar eclipse occurring it is impossible to make observations on those stars, because the glare of the daytime Sun overwhelms their effect. In that particular year there was an eclipse on that date, and preparations for making the observations were begun two years earlier.

The Astronomer Royal of the time, Sir Frank Dyson (1869–1939), was strongly in favour of organizing expeditions to observe this eclipse, and he wanted Eddington to play a leading role because in 1912 he had already led a successful expedition to Brazil to observe an eclipse of the Sun. There was, however, a difficulty. By 1917, when plans were being initiated, conscription had been introduced in Britain, and all the men with the right qualifications were eligible for the draft. Eddington was in this category; he was 34, unmarried, and physically fit. He was, however, a devout Quaker and an uncompromising conscientious objector to military service. This presented a problem not only to Eddington but to the scientific community. A brilliant young physicist, Henry Moseley (1887–1915), had volunteered for military service and at the age of only 28 had been killed at Gallipoli in 1915. It was becoming clear that scientific work was essential to Britain's success in the war; scientists were in short supply and the scientific community was pressing for exemption for qualified scientists in order for them to be available for scientific service. Eddington was already recognized as another scientist of the first rank, and a group of distinguished scientists pressed the Home Office to give him an exemption. If this had not been granted to him he might have been sentenced to jail, and at least would have been obliged to forsake scientific work and undertake work in agriculture or in industry.

Eventually the Home Office agreed, and sent Eddington a letter to sign and return. Eddington, however, was a scrupulous man, and added a note to the effect that if he were not deferred for scientific reasons he would apply for exemption on conscientious grounds. This produced a bureaucratic problem, and some of the scientists who had supported Eddington were annoyed with him for creating what they thought to be an unnecessary difficulty. The result was a further round of discussions, with the upshot that Eddington was deferred provided that he would lead the expedition—which of course is what he wanted to do.

Two sites were chosen for the astronomical expedition; one was the island of Principe, off the African coast, just north of the Congo, and it was to there that Eddington himself went to lead a Cambridge team. To reduce the odds of failure due to bad weather a second British expedition went to Sobral in northern Brazil and used a somewhat different photographic technique; it was headed by the astronomer C. R. Davidson. It was just as well that they went there, because it rained at the African site and only two satisfactory plates were obtained. Eddington arranged for these to be developed and examined on the spot, before they were transported back home. Unfortunately, during their transportation from Africa back to Cambridge the plates suffered from heat and humidity and were difficult to analyse further. As a result it was not until November that it could be announced to the world that the observed bending of the light agreed with Einstein's predictions to well within the experimental error.

No scientific experiment has ever caused so much excitement, not only in the scientific world but among the public. The official announcement was made on 6 November 1919 at a joint meeting of the Royal Astronomical Society and the Royal Society at Burlington House, Piccadilly. Dyson spoke first, followed by Eddington, and the President of the Royal Society then summed up by saying that 'This is the most important result obtained in connection with the theory of gravitation since Newton's day.'

The excitement rapidly spread to the general public. After the strain of the First World War the public had lost faith in political leaders and generals, and were looking for new heroes. The time was particularly ripe for an event that transcended blind nationalism. Here was an important scientific theory proposed by a

German-born scientist being confirmed by Englishmen so soon after a great war between their two countries. Einstein instantly became the best known media celebrity on the planet. A headline in *The Times* of 7 November 1919 announced:

> Revolution in Science: New Theory of the Universe: Newtonian Ideas Overthrown

There followed a detailed account of the Burlington House proceedings. Headlines in the *New York Times* for 10 November 1919 screamed:

> Light All Askew in the Heavens: Men of Science More or Less Agog Over Results of Eclipse Observations

and

> Einstein Theory Triumphs: Stars Not Where They Seemed or Were Calculated to Be, but Nobody Need Worry

The *New York Times* detailed one of its reporters, Henry Crouch, to cover the news of the event in England but he was their golf reporter and had little idea of what science was all about. He did have a capacity for stirring up interest and was not inhibited about taking liberties with the truth. One of the headlines for his stories ran:

> A Book for 12 Wise Men: No More in All the World Could Comprehend it, Said Einstein When His Daring Publishers Accepted It

This headline had less relationship with reality than most, which is saying a lot. Einstein was not writing a book at the time, and was dealing with no daring or even cowardly publishers. He would never have said that only 12 people could comprehend the theory of relativity, since he knew that by this time many physicists and astronomers understood it well.

It may be that it was this headline that started the idea among the public that Einstein's theory was incomprehensible even to most scientists. A story about Eddington dates from about that time. Someone said to him that he had been told that including Einstein himself there were only three people in the world who understood his theory. Eddington did not at once reply, and was asked if he agreed with the statement. 'Well,' he said, 'I've been trying to think who the third person could possibly be.' It is most unlikely that Eddington ever said any such thing, but the story is such a nice one that I believe it whether it is true or not.

Einstein himself was not impressed by all the fame that had come to him, and resisted it as much as he could. Two weeks after the public announcement of the confirmation of his theory he wrote a letter to *The Times* in which he commented rather sardonically that now that his theory had been vindicated the Germans were proudly calling him a German man of science and the English were calling him a Swiss Jew. If he had been shown to be wrong, he said, the Germans would have called him a Swiss Jew and the English would have called him a German man of science.

NINE

Energy and the universe

> It does not at present look as though Nature had designed the universe primarily for life; the normal star and the normal nebula have nothing to do with life except making it impossible. Life is the end of a chain of by-products; it seems to be the accident, and life-destroying radiation the essential.
>
> Sir James Jeans, *The Wider Aspects of Cosmogony*, 1928

At the beginning of this book we discussed the four basic types of energy that we meet in the fields of mechanics and thermodynamics: potential energy, kinetic energy, work, and heat. We often find it convenient to consider other types of energy, recognizing that strictly speaking they are not distinct forms of energy. Chemists, for example, talk about what they call *chemical energy*. When most chemical reactions occur, heat is liberated so that the system gets warmer. An extreme case of this is when a chemical reaction causes an explosion. The surroundings get hot, and in addition some material damage is generally done; matter has been displaced, which means that work has been done. It is convenient to say that chemical energy was stored in the chemical system and is released while the explosion is taking place. If we investigate these systems carefully we find that there is never any overall loss or gain of energy in a chemical process; the chemical energy has been converted into other kinds of energy, and this of course is a necessary consequence of the first law of thermodynamics. We are all familiar with this in everyday life. If we heat our homes with oil, our furnace is allowing the oil to undergo a chemical reaction (burning, which chemists call oxidation), and the energy is released as heat. What we call chemical energy results from the fact that there is a change in potential and kinetic energy when a chemical reaction occurs, owing to the reorganization of the nuclei and electrons.

Similarly, we often speak of *electrical energy*, although this again is not really a distinct form of energy. When electricity is stored in any way, such as in a capacitor, energy is being stored, and it can be released, perhaps with the production of an electric discharge. When a current flows there is movement of electrons along conductors such as metal wires, and if batteries or cells are involved there is also a movement of positive or negative particles which we call *ions*. There is energy associated with these movements, and we call this electrical energy. It can be converted for example into heat, as we know if we heat something on an electric stove. It will be remembered from Chapter 2 that Joule showed by means of careful experiments that there is an exact relationship between the electrical energy and the heat it produces.

We can also speak of *light energy*, and we saw in Chapter 7 that when light arrives at its destination it behaves as if it were composed of bundles of energy in the form of photons. When light interacts with an atom it may transfer energy to it, and an electron may move into another state where it is further from the nucleus and therefore has more potential energy. At the same time a photon is annihilated. A light quantum (a photon) of sufficient energy may actually eject an electron from an atom and be annihilated in the process (the *photoelectric effect*). The light energy from the Sun is the fundamental source of all renewable energy on the Earth. Some of it warms our Earth directly, while some of it produces the climatic effects that result in waterfalls and winds that can be harnessed as sources of energy.

When we are considering the fundamental particles of nature it is instead more useful to think in terms of four fundamental forces of nature. We may recall from Chapter 2 that a force results from a changing potential energy, so that associated with every force there is a corresponding energy. The four fundamental types of force, in order of increasing strength, are:

1. The force of *gravity*
2. The *weak nuclear* force
3. The *electrical* force—more correctly the *electromagnetic* force
4. The *strong nuclear* force.

The gravitational force is by far the weakest of them all. If we set the strength of the gravitational force as 1, in very approximate terms the strength of the weak force is 10^{25}, that of the electric force is 10^{37}, and that of the strong force is 10^{39}. The first and third of these we have already met and have discussed in a little detail. They resemble one another in obeying the inverse square law, but they differ in an important respect. The gravitational force is always attractive, while the electromagnetic force is attractive only when the charges are of opposite sign (e.g. for an atomic nucleus and an electron); it is repulsive when they are of the same sign. We can appreciate the weakness of the gravitational force compared with the electromagnetic force by noting that a nail can easily be picked up by a small magnet, even though the nail is being attracted by the enormous mass of the Earth. An apple can be held to a tree by a thin stalk, which is strong enough to counterbalance the force of gravity.

When scientists do quantum-mechanical calculations on atoms, involving the electrical forces between atomic nuclei and electrons, they can completely ignore the gravitational forces. Gravity is, however, of great importance when we are considering interactions between large bodies like stars and planets, which are electrically neutral and therefore exert no electrical forces.

The strong and weak nuclear forces, which we are now meeting for the first time in this book, are concerned with atomic nuclei. The strong force is concerned with binding protons and neutrons together in the nuclei. This interaction has only a short range, its strength falling off more rapidly than the electrostatic repulsions between protons. In atomic nuclei the strong force is about a hundred times stronger than the repulsions between the protons; this explains how it is possible for the nucleus of the uranium atom (the largest reasonably stable nucleus) to contain as many as 92 protons, packed closely together even though they repel one another strongly. There are also well over a hundred neutrons, and the strong forces provide some stability to the nucleus.

The weak nuclear force is concerned in subtle ways with radioactive decay. Its strength is about a 10-million-millionth (10^{-13}) that of the strong force. It is nevertheless much stronger than gravity—about 10^{25} times as strong. The weak interaction is what is involved in the conversion of a neutron into a proton, with the emission of an electron.

There are many aspects of the energy of the universe that we might consider, and in this chapter we will look at only a few of them. One of them is the heating of the Earth, which is now fairly well understood but for a long time caused much confusion and controversy. The trouble all started when geologists and astronomers began to determine the age of the Earth. We now know that its age is more than four and a half billion years, but until the end of the nineteenth century many investigators were arguing about ages that are much less than that. Until about 1750, in fact, most of them accepted the conclusion to which the Old Testament seemed to lead, namely that the Earth is only a few thousand years old.

Geologists, however, began to realize that such short periods were inadequate to explain their observations. Many millions of years seemed to be necessary rather than a few thousand. Strong evidence for these much longer ages was obtained by the Scottish geologist James Hutton (1726–1797). Having first studied medicine and law but never practising either, he became fascinated by chemistry and more particularly by geology, and after making a fortune out of inventing a method of manufacturing the chemical *sal ammoniac* (ammonium chloride) he settled in Edinburgh in 1768 and devoted the rest of his life to geological and other scientific investigations.

Many of his geological observations were made in Scotland, and he formulated general principles that are now widely accepted. He placed special emphasis on the role of rivers in excavating valleys and in depositing dissolved and suspended material. Sediment carried along by rivers would be washed into the sea and accumulate in deposits, which might form new rocks by the action of heat coming from the interior of the Earth. Perhaps under the influence of his friend James Watt, he regarded the Earth as a gigantic steam engine, the subterranean heat of the Earth creating upheavals from time to time, pushing up mountain ranges and twisting geological strata. He was the first to point out that the heat of the Earth's interior could explain how sedimentary rocks laid down in water could later be fused into other types of rocks such as granites and flints.

He saw no way to avoid the conclusion that many millions of years, perhaps some hundreds of millions, would be needed for mountains to be formed and be eroded. He pointed out, for example, that some of the Roman roads, laid down in Britain and in other parts of Europe nearly 2000 years earlier, were still clearly visible, little affected by erosion. Consequently, the time taken for the Earth to have been carved into its present form must have been vastly longer than the 6000 years that was popularly believed at the time. He did not commit himself to any specific time, but in 1788 he wrote: 'The result, therefore, of our present enquiry is that we find no vestige of a beginning—no prospect of an end.'

Hutton presented his work in a large book, *Theory of the Earth, with Proofs and Illustrations*, which was first published in 1785 and was expanded over the next decade. At first there was much opposition to his ideas, and for a time most geologists accepted the idea of *catastrophism*, believing that major catastrophes, like the biblical flood, were necessary to explain the Earth's development.

Another who was convinced that longer ages were required was Charles Darwin (1809–1882), who made many geological as well as biological observations during his famous voyage on HMS *Beagle* in the 1830s, the voyage that led to his theory of evolution. He too concluded that geological events must have been taking place for hundreds of millions of years. He considered, for example, a wide valley known as the Weald of Kent in south-east England, near to where he later lived for many years with his family. Darwin concluded that this valley arose from the encroachment by the sea on the line of chalk cliffs in the south of Kent. He estimated that the process would occur at the rate of one inch in a century, and from the configuration of the valley he concluded that it had been formed 300 million years ago. Darwin realized that the Earth itself may have been formed much earlier.

The distinguished geologist Charles Lyell (1797–1875) also made estimates from the same kind of evidence. His estimates, and those of many other contemporary geologists, were also in the hundreds of millions of years, in agreement with Darwin.

These estimates, although we now know them to be still too low, by a factor of at least 10, were strongly criticized by those who believed that the estimates from the scriptures were to be taken literally. There was also strong dissent from an unexpected quarter, namely from the physicist William Thomson (later Lord Kelvin) whom we have already met several times in this book. He calculated the time it would take for the Earth to cool from the temperature it had when first formed in a molten state to its present temperature. He took into careful consideration the luminosity of the Sun, the expected rate of cooling of the Earth, the formation of the Earth's solid crust, and even such relatively minor matters as the effect of lunar tides on the rate of rotation of the Earth. His initial conclusion, put forward in 1862, was that the Earth could not have solidified more than 100 million years ago. In later publications he proposed even shorter times, a few tens of millions of years. As to Darwin's estimate of 300 million years, Kelvin thought it impossibly high, since he concluded that the Earth could not even have been solid at that time; the presence of surface water, and the formation of valleys, was thus in his view impossible.

In putting forward these conclusions Kelvin displeased almost everyone, since his estimate was still much more than suggested in religious writings, but less than most geologists believed to be correct. His opinion was particularly disturbing since, in view of his work on the two laws of thermodynamics, he was recognized as a great authority on heat. By and large the geologists were sufficiently sure of their own conclusions that they ignored Kelvin's estimate, assuming that there was another source of energy in the Earth that kept it warm for much longer periods. Kelvin himself had mentioned this possibility as early as 1862, referring to 'sources of heat—now unknown to us—... in the great storehouse of creation'. He probably intended to be sarcastic in suggesting such sources of heat, believing them to be absurdly unlikely, and it is ironic that his remark proved in the end to be correct.

After the discovery of radioactivity in 1897 it soon became apparent that there was a vast store of radioactive substances in the Earth. Radioactive disintegrations are accompanied by emission of heat, which helped to keep the Earth warm for a longer period. The heat produced inside radium metal makes it feel warm to the touch. In 1903, the year in which Marie and Pierre Curie shared the Nobel Prize with Henri Becquerel for their work on radioactivity, Pierre Curie and his assistant Albert Laborde made measurements of the amount of heat produced by radium as it emits radiation. They found that 1 gram of pure radium emits in an hour enough heat to raise 1.3 grams of water from the freezing point of water to its boiling point. Put differently, radium releases enough heat to melt its own weight of ice in 1 hour. In the same year Ernest Rutherford, working at McGill University with a Canadian graduate student Howard Barnes, showed that the amount of heat produced during radioactive disintegration depends on the number of α-particles (helium nuclei) emitted. These particles collide with the nuclei in the nearby material, giving up their kinetic energy as heat.

When estimates were made of the heating effect of the radioactive material estimated to be present in the Earth, they were found to be consistent with the longer periods that the geological evidence required for the age of the Earth. Kelvin's upper limit of 100 million years for the age of the Earth is thus far too low; it is more like a few billion years.

The discovery of radioactive substances was important not only in giving an explanation for the slow rate of cooling of the Earth, but also in providing a means of measuring the ages of rocks. By the technique known as *radiometric* or *radioactive* or *radiochemical dating* it has been possible to obtain reliable values for the ages of a wide variety of rocks. Since radioactive isotopes are used in this technique, this field is often referred to as *isotope geology*. It is now a highly developed one and has led to a coherent picture of the ages of various geological events. As a result, there is overwhelmingly strong circumstantial evidence from isotope geology that the Earth is at least 4.5 billion years old.

One application of the technique is to determine the amounts of helium present in uranium minerals, and this was used by Ernest Rutherford soon after the discovery of radioactivity. The basis of the method is as follows. The isotope present in largest proportion (about 99.3 per cent) in natural uranium is uranium-238, which disintegrates as follows:

$$^{238}_{92}\text{U} \rightarrow {}^{4}_{2}\text{He (an } \alpha\text{-particle)} + {}^{234}_{90}\text{Th}.$$

The half-life of the process is about 4.5 billion years, which makes it particularly convenient for measuring the ages of rocks. Estimates of the age of a rock can be based on the fact that the helium gas produced in this disintegration is chemically inert and stable. Helium easily leaks out of molten rock, but after the rock has cooled and solidified this gas remained trapped. A determination of the strength of the radioactivity in a sample of rock, which gives the amount of uranium present, and the amount of helium trapped, will therefore, by a simple calculation, tell us how much time has elapsed since the rock solidified.

Rutherford obtained ages of about 500 million years for the particular rock samples he used, and thus proved conclusively that Kelvin was wrong. In 1904 Rutherford presented a lecture at the Royal Institution on the age of the Earth, and on the production of heat by radioactive substances. When he entered the lecture room he was somewhat disconcerted by the fact that Kelvin, then aged 80, was in the audience, but was relieved when he drifted off to sleep. Rutherford reported later that when he began to speak about the age of the Earth he saw 'the old bird sit up, open an eye and cock a baleful glance' at him. With great presence of mind Rutherford modified the thrust of his lecture somewhat, emphasizing that in 1862 Kelvin had referred to 'sources of heat now unknown to us', and that he had qualified his low estimate of the age of the Earth by writing 'provided no new source of heat is discovered'. Thus, said Rutherford with admirable tact, 'the audience must admire the foresight, almost amounting to prophesy, which had made Lord Kelvin so qualify his calculations.' Rutherford was then relieved to see that 'the old boy beamed' on him. Some of Rutherford's later work was done with the American chemist Bertram Borden Boltwood (1870–1927) who visited Manchester in 1909–1910 and measured the ages of many rocks. Boltwood, who was professor of radiochemistry at Yale, later took the technique a stage further by looking at all of the products of uranium decay, not just helium.

Much work in *radioactive dating* has been done, and the conclusion is that the Earth has existed in its present solid state for about 4.5 billion years. Some rocks are indeed formed at later times, but their age has never been found to be less than several hundred million years. The estimates of the geologists were therefore shown, by entirely independent measurements, to have been by no means too high.

Later the British geologist Arthur Holmes (1890–1965) made great contributions to the determination of geological times. In 1913, when he was only 23 years old, he published a book, *The Age of the Earth*, which soon established itself as a scientific classic. It began with the comment:

> It is perhaps a little indelicate to ask our Mother Earth her age, but Science acknowledges no shame and from time to time has boldly attempted to wrest from her a secret which is proverbially well guarded.

The book gives a masterly account of the results of the radioactive work and of how they related to the more classical studies on rates of erosion and sedimentation. Since that time the radioactive work has been greatly extended, without requiring any important changes to Holmes's early conclusions.

Some of what we know today about the Earth is obtained by monitoring seismic waves, produced when earthquakes occur. They tell us that the Earth has a solid inner core of radius about 1600 km. This is surrounded by a liquid outer core of thickness about 1800 km. The whole core is dense and is rich in iron, with a temperature of about 5000 °C. This high temperature is a result of the way the Earth was formed, as a hot ball made up from smaller masses that stuck together when the Solar System was first created. Once a crust had formed round the surface of the ball it insulated the interior, trapping the heat inside so that it escaped only slowly. This occurred over 4.5 billion years ago. Even with the heat trapped inside the Earth it could never have remained as hot as it is, as Kelvin recognized, without the infusion of additional heat from the radioactive isotopes. Even this source of heat will be used up in another 10 billion years, and by that time the Sun itself will have died so that the Earth will have become incapable of sustaining life.

The energetics of the stars, including our Sun, were also a puzzle until the importance of nuclear processes had been recognized. More, however, is involved than the radioactive disintegrations that help to keep the Earth warm. Since high temperatures exist in the stars, fission and fusion processes play the predominant role. Before considering these matters we should look briefly at some of the procedures that are used for obtaining information about stars. All of the stars except the Sun are visible to us as mere points of light, and for most of them the light takes hundreds, thousands, or millions of years for the light to reach us. In spite of this astronomers have been able to discover a great deal about their inner workings. Several techniques are used to gain information about the stars. Reliable estimates can now be made of their distances from us, and we have a good idea of the ages of many of them. The mass of the Sun can be determined precisely, from the way its gravity determines the orbits of its several planets. The masses of stars that have companions—those that form binary systems—can also be determined, but less precisely, from the way they move round one another.

Spectroscopic measurements on stars provide a variety of information: chemical composition, recession velocities of stars, surface temperatures, and energy emitted. Since every chemical element has its own unique set of spectroscopic fingerprints, the elements present in the stars can be identified without question. The amounts of the elements can also be determined. As we will discuss in more detail a little later, spectroscopists have been able to show that our Sun is composed of about 70 per cent by weight of hydrogen, 28 per cent of helium, and only about 2 per cent of all of the heavier elements. This is quite similar to the composition of the universe at large, which is something like 75 per cent hydrogen, 25 per cent helium, and less than 1 per cent of all of the other elements.

Many elements were discovered for the first time from their spectra, and helium provides a particularly interesting example. The story is a somewhat involved one, and in brief is as follows. On 18 August 1868, a total eclipse of the Sun was visible in India, and a number of scientists went there to make observations. One who examined photographs of the spectra was Joseph Norman Lockyer (1836–1920) who although a civil servant at the British War Office had already in his spare time done valuable work in astronomical spectroscopy. For help with examining the spectra he turned to the distinguished organic chemist Edward Frankland (1825–1899).

Lockyer was particularly interested in a so-called D_3 line in the yellow region of solar spectra. It was known that the well-known sodium D line was in fact two lines close together, called the D_1 and D_2 lines. The D_3 line could not be obtained from any substance available in the laboratory, and Lockyer boldly suggested that it was caused by a new element, found in the Sun but apparently not on Earth. He gave this new element the name helium, from a Greek word meaning the sun.

Much later, in 1895, William Ramsay (1852–1916), who with Lord Rayleigh had already discovered the element argon, began to investigate the gas produced by a mineral called cleveite. On examining its spectrum he found a line which he remembered, from a lecture by Lockyer that he had attended many years previously, to have been called the D_3 line and to have been tentatively identified as relating to a new element. He sent a sample to Lockyer, who was then director of the Solar Physics Laboratory in South Kensington, and who turned over his whole laboratory to the study of the material. The existence of helium was soon confirmed. Ramsay was knighted in 1902, and in 1904 received the Nobel Prize in Chemistry for his discovery of inert gases and his investigations of their chemical and physical properties.

For his contribution to the discovery of helium and other scientific contributions Lockyer was knighted in 1897. He made many contributions to spectroscopy and astronomy, but perhaps his most important and lasting achievement was his founding in 1869 of the journal *Nature*, which he edited for the first 50 years of its existence, and which continues to play a distinguished role in the communication of science.

Spectra play other important roles in spectroscopic investigations. Through what is called the Doppler effect they allow us to determine accurately the speed with which the stars are moving away from us, which nearly all of them are doing. We saw in Chapters 5 and 6, in connection with the distribution of the energies of molecules, that the temperature influences the form of the distribution curve. As a result the spectrum of a star allows the temperature of its surface to be estimated.

During the nineteenth century scientists began to think about the physics of the stars and about their ages. Particular attention was paid to the Sun, for that is the star we know most about. Since geologists had been concluding that the Earth had existed for at least many hundreds of millions of years, it was obvious that the Sun must be a good deal older. This, however, presented a problem, because it was obvious that no ordinary chemical reaction could keep the Sun hot for anything like that length of time. It the Sun were a solid lump of coal, and was burning in pure oxygen sufficiently vigorously as to generate the amount of heat that the Sun is known to do, it would have burnt to a crisp in about 1500 years.

Two distinguished physicists whom we met in earlier chapters, William Thomson (Kelvin) and Hermann von Helmholtz, applied themselves to this problem. Their suggestion, which turned out to be only part of the answer, was that gravity played an important role. If the Sun started out as a thin cloud of gas the force of gravity would make it more and more dense and compact. Its potential energy would therefore decrease, and by the first law of thermodynamics there would be evolution of heat. When Kelvin and Helmholtz made the necessary calculations they obtained ages of 100 million and 25 million years respectively. Since Kelvin had estimated that the age of the Earth was somewhat less than 100 million years, he though that his estimates were satisfactorily self-consistent. The geological evidence, however, strongly favoured much longer ages. When radioactivity was discovered it was realized that it provided a way for the Earth to have survived for much longer, and as we shall see nuclear processes provided a way out of the dilemma as far as the Sun and other stars were concerned.

At first it was thought that radioactive disintegrations contributed in an important way to the heating of the Sun. In 1903 the English astronomer William Watson calculated that if there were 3.6 grams of pure radium in each cubic metre of the Sun's volume, the heat evolved in the radioactive decay would be enough to supply all the heat radiated by the Sun. Similar ideas were developed by the astronomer George Darwin, one of the sons of Charles Darwin. This idea, however, was inadequate, because there is not enough radioactive material present. There is an interesting story behind this statement. Much of our understanding of the composition of the Sun came originally from the work of Cecilia Payne-Gaposchkin (1900–1979). Born Cecilia Helena Payne in Wendover, Buckinghamshire, she showed great aptitude for science and mathematics at an early age. While still at school she tested the efficacy of prayer by a controlled test: she divided her examinations into two groups and prayed for success in one group only. Since she actually got better marks in the prayerless group she became, and remained, a devout agnostic. In 1919 she went to Newnham College, Cambridge—significantly an agnostic foundation—where she attended lectures by Arthur Eddington, some of them on his confirmation of relativity theory which we discussed in the last chapter. These lectures particularly inspired her to become an astronomer.

After leaving Cambridge she was awarded a fellowship to Radcliffe College, to do research at the Harvard College Observatory where she spent her entire career. For her PhD degree at Harvard she studied stellar spectra, and was able to deduce, for the first time, reliable pressures, temperatures, and compositions for a variety of stars. The important conclusion at which she arrived was that all of the stars have remarkably similar compositions, consisting predominantly of hydrogen and helium, with only small amounts of the heavier elements. Today we know her conclusions to be correct, the composition of the stars being much the same as that of the universe as a whole, and very different from the composition of the Earth, which contains much more of the heavier elements. In the 1920s, however, it was firmly believed that the composition of the Sun was much like that of the Earth (which was thought to have split off from the Sun), and Payne's conclusion was hotly disputed. Her research director, the famous astronomer Harlow Shapley (1885–1972), was sceptical, and her PhD examiner, the equally famous Henry Norris Russell (1877–1957), insisted that she add an important qualification to her thesis before he would accept it. She was required to write, with regard to hydrogen and helium, that 'the enormous abundance derived for these elements in the stellar atmosphere is almost certainly not real'. It later became clear that it certainly was real. Cecilia Payne obtained her PhD degree in 1925, and the conclusions from her work appeared in a book, *Stellar Atmospheres*, which was published in the same year. Her later career was distinguished; she did much more research in astronomy, and in 1956 became the first woman professor at Harvard.

It is thus now certain that the Sun and stars do not contain enough radioactive material to keep them warm for so long. The correct answer to the problem was provided by the great astronomer and cosmologist Arthur Stanley Eddington (1882–1944; Fig. 25), whom we met in the last chapter in connection with his

Fig. 25 Sir Arthur Eddington (1882–1944), British astronomer and often regarded as the founder of the science of astrophysics. He led a team of scientists who performed a crucial and successful test of Einstein's theory of relativity. He was also the first to formulate a satisfactory theory of the constitution of the stars.

verification of Einstein's theory of relativity. He was born in 1882 in Kendall, Westmoreland, and first took a degree from Owens College, Manchester, which is now merged into the University of Manchester. From there he went to Trinity College, Cambridge, and in 1904 graduated as Senior Wrangler in the Mathematical Tripos. He was elected a Fellow of Trinity but soon left Cambridge to take up the position of chief assistant at the Royal Observatory in Greenwich. In 1913, at the age of only 31 he was appointed Plumian Professor of Astronomy and Natural Philosophy (which means physics) at Cambridge. In the following year he also became director of the University Observatory.

He became particularly interested in the constitutions of stars in 1916. Four years later, at the annual meeting in Cardiff of the BAAS, he gave a lecture which is considered to have constituted the birth of the science of astrophysics. As we saw in Chapter 8, Eddington at the time had become a great expert on the theory of relativity and appreciated the implications of Einstein's famous formula $E = mc^2$. In 1920 no fission or fusion process had yet been discovered, but Eddington had already realized that they were possible, and that under the conditions present in the stars they would occur and contribute to extending the lives of the stars. To explain how the Sun could have stayed hot for as long as it appeared to have done he suggested that an important role was played by the conversion of hydrogen nuclei into helium, although at the time the details of how this could happen were still not worked out. It was known that the universe consisted largely of hydrogen with about a third as much helium, and it was natural to conclude that there must be some mechanism by which helium atoms were being produced from hydrogen atoms. From Einstein's equation Eddington was able to deduce that when four hydrogen nuclei come together to form a helium nucleus much energy is liberated, the mass balance (using modern values) being as follows:

$$\begin{aligned} \text{mass of four protons} &= 4 \times 1.00794 = 4.0318 \text{ au} \\ \text{mass of one helium nucleus} &= 4.0026 \text{ au} \\ \text{loss of mass} &= 0.0292 \text{ au.} \end{aligned}$$

Since 1 au is equivalent to 931.5 MeV (Chapter 8) the energy released is 27.20 MeV.

From such data, and from the amount of energy that the Sun is actually radiating, Eddington was able to estimate that the brightness of the Sun can be maintained if about 4 million tonnes of the Sun's matter are being converted into energy every second, by the formation of helium atoms from hydrogen atoms. That seems a great deal, but it is only a tiny fraction of the mass of the Sun, which is roughly 2×10^{27} tonnes. To convert that amount of matter into energy the mass of hydrogen that has to be converted is about 600 million tonnes of hydrogen per second, or about 2×10^{16} tonnes of hydrogen per year. In 10 billion (10^{10}) years it has therefore consumed about 2×10^{26} tonnes, or 10 per cent of the entire mass of the Sun. This conclusion is quite consistent with the estimate that the Sun has an age of about 5 billion years, and it will probably survive in more or less its present form for another 5 billion years.

Since the work of Kelvin and Helmholtz, astronomers had been aware of the fact that gravitational energy would be converted into heat. Eddington's calculations, however, showed that such a process could only keep the Sun hot for a few tens of millions of years, not for the billions of years required by the geologists (4.5 billion years as we now know). In his address to the BAAS in 1920 Eddington brought out this point forcibly and convincingly, emphasizing that not just radioactivity was involved: nuclear processes had to be playing a dominant role. He thus deduced the occurrence of these nuclear processes long before any of the fission or fusion reactions had been discovered. He even made a prediction about nuclear energy, saying that 'we sometimes dream that man will one day learn to release it and use it for his service.'

Over the next few years Eddington pursued this line of investigation, showing that from the basic laws of physics one could make important deductions about the nature of stars. If a star is at equilibrium, neither expanding nor contracting, the inward pull of gravity must be balanced by the outward pressure. This is partly the radiation pressure resulting from the radiation emitted by the hot gases, and partly the pressure of the gas that results from the kinetic energy of the motions of the atoms and molecules present. Knowing the mass of a star and with certain other information we can calculate from the laws of physics what its size and temperature must be and how much energy it radiates. For many stars it is thus possible to be able to deduce an energy balance. In particular, we can infer the amount of energy that is radiated from the surface of the star. Since this is something that can be measured directly we can therefore see whether the model proposed for the star is self-consistent.

At the temperatures that exist inside the stars the electrons are stripped off the atoms, and there exists a so-called *plasma* of charged particles. The fast-moving particles emit radiation, and there is a resulting radiation pressure. If a ball of gas has more than a certain critical amount of mass, the fast-moving particles generate so much radiation pressure that the gas is blown apart. By 1920 Eddington had worked out a theory of this, and he realized that there are three possible fates for a hot ball of gas. If it is too small a plasma cannot form and it will become a cool mass in which there is little radiation pressure, so that the inward gravitational attraction is balanced by only the gas pressure; the planet Saturn is a body of this kind. If it is somewhat bigger it can become a stable glowing star like our Sun, with the gas pressure plus the radiation pressure balancing the gravitational pressure. If it is bigger still it will be unstable, shining brilliantly before being blown apart by the enormous radiation pressure. By reasoning in this way Eddington was able to work out the range of masses over which stars can exist. Even though today we know much more about the details of what stars are made of, and of what nuclear processes can occur, modern estimates of the range of masses are not too different from Eddington's.

Another important conclusion reached by Eddington is that all stars, whatever their mass and brightness, must have much the same temperatures at their centres. His estimate of that temperature was a little higher than the accepted modern value of 15 to 20 million degrees. The reason that all stars have a similar internal temperature is *negative feedback*. This is what is involved in the thermostats that control the temperatures of our homes. If the temperature gets too high the thermostat turns the furnace or boiler off and the house cools down; when it gets too cold the furnace or boiler goes on again. It is like that in a star. If it shrinks a little the increased gravitational attractions will cause a release of energy so that the star will get hotter. That causes the star to expand, until equilibrium is restored. If on the other hand the star expands, the lowering of the gravitational attractions will cause cooling, which causes it to contract again. A star thus has a built-in thermostat which keeps its core at a fairly constant temperature. This temperature is high enough for nuclear reactions to occur, and they enable the star to live much longer that it otherwise could.

Eddington summed up these basic conclusions in his great classic *The Internal Constitution of the Stars*, which appeared in 1926. It is remarkable that he was able to reach such a detailed understanding of the energetics and other features of the stars before much was known of nuclear reactions, and when quantum physics was still in its infancy. For quite a few years Eddington's ideas were dismissed by many physicists, but later it was realized that he had been essentially right all along.

We should now consider in more detail the role played by the nuclear reactions in the stars. At first sight we would think that a nuclear reaction occurring at the heart of a star would increase its temperature. However, because of feedback this is not what happens. Without the nuclear reactions, which are producing energy and therefore provide an outward pressure to resist the inward pull of gravity, the star would shrink and therefore get hotter still. The nuclear reactions thus have the effect of preventing the collapse of the star as a result of gravity, and therefore of prolonging its life.

One of the objections that physicists raised to Eddington's ideas about the interiors of stars was that even at the enormous temperatures in the stars the nuclear reactions could not occur sufficiently rapidly. One of the simplest of the nuclear reactions is the fusion of two deuterium nuclei to give a helium nucleus. Deuterons, however, being positively charged repel each other strongly, and Eddington's critics objected that even at the enormous temperatures in the stars the deuterons would not have enough energy for them to come together and undergo this process; much higher temperatures would be required for them to do so. Eddington stuck to his guns and pointed out that helium does exist to a considerable extent in the universe. In his own words, 'the helium that we handle must have been put together at some time and some place.' He then added, perhaps tongue in cheek, a comment that is often quoted: 'We do not argue with the critic who urges that the stars are not hot enough for this purpose; we tell him to go and find a hotter place.' This is usually interpreted as Eddington's polite way of telling his critics to 'go to Hell'.

Soon after Eddington's book appeared the answer to the apparent dilemma was suggested by the Russian–American physicist George Gamow (1904–1968), who made many important contributions to the understanding of nuclear processes in the stars. Born in Odessa, he was educated at the University of Leningrad, where he was later a professor of physics. After research at both Göttingen and Cambridge he moved to the United States and from 1934 to 1955 was professor of physics at George Washington University in Washington, DC. After that he spent his final years at the University of Colorado. He was a large enthusiastic man with a great sense of humour. He created the character of Mr Tompkins and in a number of popular books used him to communicate to the general public an understanding of relativity, particle physics, and quantum mechanics. I met Gamow a few times at scientific meetings, and was always impressed by the fact that he spoke many languages with great fluency. The odd thing, however, was that each language as he spoke it sounded exactly the same.

In 1928, just two years after Eddington's book appeared, Gamow applied the new quantum mechanics to nuclear processes. Heisenberg's principle of uncertainty had been proposed in 1926, and the new quantum mechanics allowed the possibility of what is called *quantum-mechanical tunnelling*. The idea behind this is as follows. Suppose that a vehicle needs to surmount a hill, and that a certain amount of energy will allow it to do so. If the vehicle approaches the hill with too little energy it cannot surmount it. This is certainly true for a vehicle to which classical mechanics applies, but for particles on the atomic scale the situation is different. Quantum mechanics does allow a particle of atomic size to get from one side of a hill to the other even if it does not have enough energy to surmount it. We say that the particle *tunnels* through the hill. This is the point that was made by Gamow in his 1938 publication. Two deuterons can collide and interact with one another, admittedly with a low probability, even though they do not have enough energy to surmount the energy barrier. This explanation successfully overcame the difficulty that appeared to make the nuclear processes impossible. Gamow also showed in the same investigation that in spite of the strong repulsive force, two protons can fairly easily interact with the formation of a deuterium atom. Since in addition a proton can easily combine with a neutron it follows that wherever there is ${}^{1}_{1}H$ in a star there is always a supply of deuterium, ${}^{2}_{1}H$. At the high temperatures of the stars these fuse together so rapidly to form helium that the proportion of deuterium to ordinary hydrogen remains low.

Let us now look briefly at the life history of a star, concerning ourselves particularly with the energies involved. Some of what we conclude about the stars does not come to us directly from observational evidence, but from computer simulations of what occurs in the stars, based on our knowledge of the laws of physics. Support for these deductions comes from a variety of experiments carried out on Earth, for example to determine the rates with which certain nuclear reactions occur under the conditions believed to exist in the stars.

As a result of investigations of this kind a whole body of knowledge about the stars has now been constructed. Everything that has been deduced forms such a coherent package that it is difficult to believe that it can be far from the truth. The kind of information that we can deduce about a star is in brief as follows. If we know the mass of a star and the temperature at its surface, the basic laws of physics. the relevant data, and the computer models allow us to deduce how hot the star is at its interior, and what its pressure and density must be. Then, knowing the temperature at the centre of the star and its composition we can deduce something about the nuclear reactions that must be going on in the interior. Also, from experiments carried out on Earth we know how much energy is being released in the star. This can then be compared with the amount of energy that the star actually does radiate. In this way we can confirm that our model of what is going on in the star is self-consistent.

We know that star formation is going on all the time, by the recycling of material from clouds of gas and dust in space. After a star is formed it is surrounded by clouds of material, some of it in the form of spinning discs from which planets may form. Star formation occurs when such a cloud gets squeezed from outside, perhaps by shock waves. It then starts to collapse as a result of gravitational forces, and while it collapses it fragments, forming binary stars and other structures. The initial pressure that causes the collapse is due to compression waves that move round the galaxies, perhaps as a result of an exploding star. A compression wave, of which a sound wave is an example, squeezes the gas as it passes, but leaves individual molecules of air more or less undisturbed after it has passed. The largest stars that are formed in this way have only relatively short life times (no more than a few million years rather than about 10 billion for our Sun).

A star like the Sun, however, has a size that is conducive to longevity. After it has formed, and has shrunk by gravitational attraction until its temperature has reached about 15 million degrees, it begins to operate as a nuclear fusion reactor, converting hydrogen into deuterium and then into helium. As long as hydrogen remains in the core these nuclear reactions keep the star from reacting further and getting any hotter. The lifetime of such a star, which is known as a *main-sequence star*, depends on its mass. The bigger it is the less time it spends as a main-sequence star, because it burns its fuel more vigorously to keep from contracting and getting hotter. A star 25 times more massive than the Sun spends only about 3 million years as a main-sequence star. We have seen that by contrast the Sun will last a total of about 10 billion years, and that since it has already shone for about 5 billion years it is about half-way through its life as a main-sequence star. A star half as massive as the Sun may last about 200 billion years. When all of the hydrogen in the core of a star like the Sun has been converted into helium, so that its nuclear energy supply is depleted, its core begins to shrink and generate heat as gravitational energy is released. The rise in temperature will make the outer part of the star expand and turn the star into what is called a *red giant*. When this happens some matter is blown away into space and the star swells up to almost the size of the orbit of Venus. During this time the temperature of the core will reach a temperature of about 100 million degrees, and at such temperatures helium is converted into carbon. The star then becomes stabilized as a red giant, which will survive for about a billion years after which its helium supply will be exhausted.

At the end of this phase the carbon core begins to collapse, releasing enough heat to allow hydrogen to burn for a further period, farther out from the centre of the star. During this period the star will expand to reach the present orbit of the Earth, at the same time releasing elements such as nitrogen and carbon into space. Nuclear processes are now unimportant because of lack of fuel, and eventually the inner core, largely carbon, simply cools down. The star eventually settles down as a white dwarf and will be about the present size of the Earth. Its density will therefore be enormous, since today the volume of the Sun is roughly a million times that of the Earth. The density of the white dwarf will thus be about a million times that of water; 1 cubic centimetre would have a mass of over 1 tonne. Consistent with this conclusion, no white dwarf has been observed that has more than 1.4 times the mass of the Sun. This fits in with the inference from theory that a star that has such a mass has a significantly different life history. A heavy star will go through more stages of nuclear burning, at higher and higher temperatures, and will manufacture more of the heavier elements such as carbon, iron, and still heavier elements such as uranium. What will happen in the case of a star that is more than 1.4 times as heavy as the Sun is that the gravitational compression is so great that many of the electrons present combine with protons to form neutrons, by a process that is the reverse of the beta decay of a neutron:

$$^{1}_{1}\mathrm{H} + {}^{0}_{-1}\mathrm{e}\ \text{(an electron)} \rightarrow {}^{1}_{0}\mathrm{n}.$$

In this way the star is reduced to an enormous mass of neutrons, which since they do not repel one another can be packed together closely. It thus forms what is called a *neutron star*, which is much more dense than a white dwarf. We have seen that a white dwarf with the mass of the Sun has the size of the Earth; a neutron star with 1.5 times the mass of the Sun will be much smaller, with a diameter of only about 10 kilometres. Each cubic centimetre of it would weigh about a billion tonnes. There is a limit to the size of a neutron star, since if the mass is too great the gravitational forces are large enough to cause it to shrink indefinitely and form a *black hole*. At a black hole the gravitational field is so strong that nothing can escape from it, not even light.

The energetics of what occurred at the creation of the universe is of great interest, and we can go into it only briefly. It is now generally accepted by astronomers and cosmologists that the so-called Big Bang theory gives the best interpretation of what occurred at the beginning of time. This theory was inspired for the most part by the observations by the American astronomer Edwin Powell Hubble (1889–1953; Fig. 26), who was born in Marshfield, Missouri, the son of a not very successful lawyer. He became interested in astronomy as a child. He studied science and mathematics at the University of Chicago and won a Rhodes Scholarship to Queen's College, Oxford, where he acquired an exaggerated 'Oxford manner' that remained with him all his life. At Oxford he studied law but never practised it. During his student days in Chicago and Oxford he devoted a good deal of time to basketball and boxing, and was so accomplished a boxer that he was urged to become a professional. Fortunately for astronomy he rejected that career, and in 1914 became a research student in astronomy at the Yerkes Observatory, operated

Fig. 26 Edwin Powell Hubble (1889–1953), American astronomer and cosmologist, who discovered the expansion of the universe, and whose work led to estimates of its size and age.

by the University of Chicago. There he made a particular study of the groups of stars called nebulae or galaxies, and his PhD degree in 1917 was concerned with the classification of the nebulae according to their appearance.

After service in the First World War he joined the Mount Wilson Observatory in California and soon established a great reputation. He is particularly remembered for his discovery that the vast majority of the galaxies are moving away from us, and that the more distant ones are moving away more rapidly than the nearer ones. In fact, there is a *proportional* or *linear* relationship between rate and distance: if a galaxy is twice as far from us as another one it is receding twice as fast. The speed of recession is thus related to its distance by the formula

$$\text{speed of recession} = \text{Hubble's constant} \times \text{distance}.$$

Because of the enormous distances involved for the distant galaxies, the value of the Hubble constant is difficult to obtain precisely.

The galaxies are slowing down as time goes on because of the gravitational attractions between them, but the effect is small and does not greatly affect our estimates of the age of the universe. Suppose that we imagine going back in time, which we can do by imagining that the speeds of the galaxies are all reversed. As we go back in time the ones that are far away come back at high speed, the nearer ones at lower speeds, until at zero time they are at the same place. In other words, at some time in the far distant past they must all have been in a small region of space. The obvious conclusion is that the universe was created by an enormous explosion in which, as in all explosions, some fragments were propelled much more rapidly than others. The ones that were propelled at the highest speeds have by the present time travelled the longest distances, and are at the outer extremities of the universe. The slowest ones have not gone anything like as far. From the size of the Hubble constant it is easy to calculate that the explosion must have occurred at least 12 billion years ago.

The universe is thus expanding, from highly dense matter which exploded, the fragments travelling in all directions at immense speeds. It is important to understand that the suggestion is not that we are at the centre of the universe, and that everything else is rushing away from us. A useful analogy is provided by a lump of dough, with currants in it, that we place in an oven. As it cooks, the whole thing swells, and the distance between every pair of currants increases. In the same way, as our universe expands the distance between almost every pair of galaxies increases. There are some exceptions; for example, because of the attraction of our Galaxy the neighbouring galaxy Andromeda is approaching us.

The *Big Bang* theory was one of the first theories of this expansion, and it is the only one that is now supported by almost all scientists. The expression 'Big Bang' was given to it rather scornfully by the British cosmologist Fred Hoyle (1915–2001), who supported a rival theory. In 1927 the Belgian astronomer the Abbé Georges Edouard Lemaître (1894–1966) deduced from Einstein's theory of relativity that the universe might be expanding, even before there was much evidence for such an expansion. Later, when the idea of an expanding universe had been strongly supported by the observations of Hubble and others, Lemaître developed his ideas in various directions, and it was he in particular who suggested that the universe was originally small and highly compressed. Today this is the most popular theory of the origin of the universe.

Since energy is conserved, all of the energy that is in the universe today must have been concentrated in a much smaller volume when the Big Bang occurred. The event must therefore have been accompanied by the production of incredibly high temperatures. Initial temperatures of over a billion billion degrees have been estimated, but the exact value does not matter to the argument. The point is that within a tiny fraction of a second there was a great expansion, and the temperature dropped enormously. The important temperature to keep in mind is 3000 K, because above that temperature electrons cannot become associated with atomic nuclei but freely pervaded the universe. Under these conditions, estimated to have lasted for about 500 000 years, the universe was opaque, in the sense that the photons (the particles of light) could not travel freely because of the electrons in their way. After the temperature had fallen below about 3000 K, however, the situation greatly changed. The atmosphere became transparent, and light travelled freely. Electrons became associated with nuclei, and some molecules could be formed by the uniting of atoms.

Two lines of evidence, quite different from each other, have provided overwhelmingly strong support for the Big Bang theory. The first of these, the *cosmic microwave background radiation*, was discovered by chance. In the 1960s the American astrophysicists Arno Allan Penzias (b. 1933) and Robert Woodrow

Wilson (b. 1936) of the Bell Laboratories in New Jersey were exploring the Milky Way with a radio telescope, their object being to improve communications with satellites. At a wavelength of 7 centimetres, which is in the microwave region of the spectrum, they found a greater effect than could be accounted for by any known source. They found that the radiation was equally strong from all directions, even coming from apparently empty sky. From the characteristics of the radiation they were able to establish that it was being emitted from a source that had an apparent temperature of about 3 K, that is −270 °C.

It was finally realized, and no better explanation has been suggested, that this microwave radiation must be a kind of afterglow, or 'fossil', left behind by the Big Bang. As cooling occurred, the strong forces first came into play and atomic nuclei formed. After the universe had cooled to about 3000 K the electrons became associated with protons and neutrons to form atoms, and the universe became transparent to radiation. The temperature of 3 K for the background radiation tells us that since the time when the universe became transparent until now it has expanded by a factor of 1000, causing the temperature to drop from 3000 K to 3 K. At the same time the wavelengths have increased by the same factor. For their discovery of the background radiation Penzias and Wilson shared the Nobel Prize in Physics for 1978.

More recent work has given much support to this interpretation of the microwave radiation. In November 1989 the Cosmic Background Explorer (COBE) satellite was launched into orbit above the Earth's atmosphere, and obtained results that agreed exactly with the predictions of the Big Bang theory—so well that the average temperature of the radiation can now be given more precisely as 2.736 K. This consistency of the data is so good that we can now be confident that the background radiation is indeed fossil radiation left over after the Big Bang. One refinement of the investigations has been particularly impressive. The Big Bang theory leads to the conclusion that when the universe had cooled to 3000 K there would have been density fluctuations; certain regions, where galaxies were later to form, would have been of higher density than the rest. Corresponding to these fluctuations there would be regions of space where the temperatures would be different, but only by one part in about 100 000. These tiny fluctuations over the sky were actually observed in the COBE experiments, exactly as predicted by the Big Bang theory.

The second line of evidence in favour of the Big Bang theory is that it leads to a satisfactory theory of *nucleosynthesis*—the formation of the atomic nuclei—which explains the *distribution of elements* in the universe. We have seen that the atom of each element has a nucleus with electrons orbiting around it, the number of orbiting electrons being equal to the positive charge on the nucleus, this number defining the identity of the element. The nucleus is different for each chemical element, but contains only protons and neutrons. There are about 90 reasonably stable chemical elements, but six of them play a particularly important role in our universe: hydrogen, helium, oxygen, nitrogen, carbon, and phosphorus. The status of the first two is rather remarkable, since in the universe as a whole they are by far the most abundant elements. Of all the matter in the universe, about three-quarters (by weight) is hydrogen, and one-quarter is helium. (Put differently, about 90 per cent of all the atoms in the universe are hydrogen atoms, and about 9 per cent are helium atoms; a helium atom is four times as heavy as a hydrogen atom.) All the rest of the elements together comprise only a very tiny proportion, between 1 and 2 per cent by weight.

Hydrogen is plentiful on the Earth but hardly exists at all in the uncombined state. Because of its low density any free hydrogen gas on Earth soon floats to the top of the atmosphere and disappears into space. Helium only exists in the neighbourhood of radioactive substances which produce it as they disintegrate; because of the low density of helium it also floats away into the upper atmosphere and beyond. The element helium is highly inert in the sense that it forms no compounds. Hydrogen does readily form chemical compounds, which are abundant on the Earth, particularly water, H_2O, but also in the form of many compounds that occur in fossil fuels and in all animals and plants.

A crucial test of the Big Bang theory is to see whether it can give a satisfactory explanation for the distribution of elements in the universe. The most striking feature of this is that hydrogen and helium are by far the most abundant, and that all the other elements are much less abundant. The obvious explanation is that hydrogen was formed first, and then helium, and that the rest of the elements were formed later from these two elements. The first explanation as to how helium was formed was put forward by George Gamow, whom we met earlier in this chapter. In publishing this proposal he perpetrated a joke that has become a scientific classic. The research was actually done with his student Ralph Alpher (b. 1921), but Gamow felt that to publish with Alpher alone 'seemed unfair to the Greek alphabet.' He therefore added to the paper the name of Professor Hans Bethe (b. 1906), who had already made important contributions to the theory of nuclear reactions in the stars. As a result the paper appeared in 1948 under the authorship of Alpher, Bethe, and Gamow. The fact that the date of the publication was April the first was a particular delight to Gamow. One detail about the story remains obscure to this day, since both Gamow and Bethe were reticent about it. One version is that Gamow got in touch with Bethe, sending him a draft of the paper asking if he would agree to his name being added, and that Bethe had no objection. The other is that Gamow mischievously added Bethe's name without consulting him, and that Bethe was at first not at all pleased, but good-naturedly took the matter in good part. Perhaps the truth will never be known with certainly; for my part I prefer the second version. Another dubious detail is that the words 'in absentia' were originally added after Bethe's name, but that they mysteriously disappeared before publication.

The essence of the theory is as follows. First, protons became associated with neutrons, to form deuterium (Fig. 24), and this they do readily at enormous temperatures. Deuterium atoms, however, are uncommon in our universe, much less common than helium atoms, and this is because at the high temperatures of the early universe they rapidly unite to form helium. In Chapter 8 we discussed the fusion reaction in which two deuterium nuclei come together to form a helium nucleus (two protons and two neutrons):

$${}^{2}_{1}\text{H} + {}^{2}_{1}\text{H} \rightarrow {}^{4}_{2}\text{He} + 23.7 \text{ MeV}.$$

At the high temperatures shortly after the Big Bang this process occurred readily, its driving force being the large amount of energy evolved in this fusion process. As a result, practically every deuterium nucleus formed immediately after the Big Bang was converted into a helium nucleus. We will not go into further details, but it can be deduced from the Big Bang theory and our understanding of nuclear physics just why the amount of hydrogen in the universe is almost exactly three times the amount of helium. This alone is a very convincing argument in favour of the theory.

We will conclude this chapter by mentioning briefly another elementary particle of nature, important in connection with the subject of energy. So far in this book we have concentrated on just three elementary particles: electrons, protons, and neutrons. Of these the protons and neutrons have just about the same mass, while the electrons are roughly 2000 times lighter. Much of the behaviour can be understood in terms of these particles, but over the years it has been realized that there is evidence for the existence of many other elementary particles, which are involved in a variety of processes. We will mention only one of them, the *neutrino*.

The evidence for this particle was at first indirect, or circumstantial. Experimental studies had been made of a number of radioactive processes in which there was beta decay, by which we mean that a β-particle, namely an electron, was emitted by one nucleus with the formation of another. When measurements were made of the energies of the emitted neutrons there was a surprise. It was expected that all of the electrons would be emitted with the same energy, this energy being the difference between the energy of the emitting nucleus and that of the product. Some electrons did indeed appear with a certain maximum energy corresponding to this difference, but many appeared with lower energies. How was it possible for electrons to appear with lower energies; where was the missing energy?

There were only two possibilities. Energy was not being conserved, or there was some other particle which was emitted at the same time and which carried away the missing energy, in some manner that had hitherto escaped detection. The principle of conservation of energy has been established by so many experiments over a long period of time that no one took seriously the idea that it could be violated when a radioactive process occurs. The other explanation was therefore favoured, and in 1931 the Austrian-born physicist Wolfgang Pauli (1900–1958) suggested that the energy was being carried away by a particle that has no charge and is so light (being perhaps completely without mass) that it is difficult to detect. The Italian-born physicist Enrico Fermi (1901–1954) gave this elusive particle the name *neutrino* (Italian for 'little neutral one') to the particle.

It took some time to detect neutrinos since they interact so weakly with matter. They pass through planets, our bodies, and the Sun as if they were not there.

Experimental evidence for their existence was finally obtained in 1956 by two American physicists, Frederick Reines (b. 1918) and Clyde Lorrain Cowan (1919–1974). Working with collaborators from the Los Alamos Scientific Laboratory, they used sensitive detectors placed near to a powerful reactor in which fission processes were occurring. Much work on neutrinos has subsequently been carried out, and today this field is a very active one. Neutrinos coming from outer space are sometimes detected experimentally by means of large tanks of chlorinated hydrocarbons placed in a deep mine, about 3 kilometres deep. The neutrinos interact with chlorine atoms to produce radioactive argon atoms, the number of which can be measured from the rate of their radioactive decay. The detectors sometimes present an area of 200 square metres, and may weigh thousands of tonnes.

It is estimated that every second about 70 billion (7×10^{10}) neutrinos pass through every square centimetre of our bodies. In February 1987 a supernova (an exploding star) was observed in the Large Magellanic Cloud, and it is estimated that some 10^{58} neutrinos were liberated. They were released 160000 years ago, and after spreading out a distance of 160000 light years about 3×10^{14} of them should have passed through the large underground tanks in use for their detection. Of these, just 19 were detected in Japan, and 8 in the United States.

In April 2002 the announcement was made of the results of important neutrino experiments carried out by a joint Canadian–American–British team of scientists at the Sudbury Neutrino Observatory (SNO), in Sudbury, Ontario. The detector, set up at the bottom of a mine 2039 metres below the surface, was a spherical vessel containing 1000 tonnes of ultra-pure heavy water (D_2O), surrounded by ultra-pure ordinary water which screens out radiation from other sources. It is now known that neutrinos exist in three varieties, or 'flavours', and that they change their form as they travel to the Earth. They are emitted by the Sun as electron-neutrinos, denoted by ν_e, but as they travel they may be converted into forms known as mu-neutrinos (ν_μ) and tau-neutrinos (ν_τ). Detection of the neutrinos is achieved by making use of two processes that occur. One of these, which occurs only with the electron-neutrinos, is

1. ν_e + deuterium nucleus → 2 protons + electron.

The electron reacts with heavy water with the production of photons, which are detected by photomultiplier cells supported throughout the ordinary water that surrounds the heavy water. In the second process, a neutrino in any of its three forms, denoted by ν_x, interacts with a deuterium nucleus to form a neutron which adds on to another deuterium nucleus to give a tritium nucleus with the emission of gamma (γ) rays:

2. ν_x + deuterium nucleus → proton + neutron
   proton + deuterium nucleus → tritium + γ photon.

Over a period of 308 days reliable evidence was obtained for the detection of about 2800 neutrinos. By detection of the radiation emitted, and in other ways, it proved possible to confirm that the rate of emission of neutrinos from the Sun is consistent with current theories of the processes occurring in the Sun. The latest experiments at Sudbury also showed convincingly that the neutrino has a non-zero mass, although an extremely small one. Its mass cannot be more than a millionth of the mass of the electron, which is one of the lightest of the fundamental particles.

The great importance of the SNO experiments is that they completely confirm the theories of the processes occurring in the Sun, which have been outlined in the present chapter. We can now say with some confidence that we know just how the Sun shines.

TEN

Chaos: the science of the unexpected

You can only predict things after they have happened.

Eugène Ionesco, *Le Rhinoscéros*, 1959

Scientists cannot predict the future any better than anyone else—even about their own field of research.

John Kendrew, *The Thread of Life*, 1966

Chaos, in the sense used in modern chaos theory, is something we constantly see all around us—whenever it rains, every time we turn on a tap. Some confusion about the subject has arisen from the fact that scientists are now using the word chaos in two senses that are somewhat related to one another but must be clearly distinguished. The first and original sense corresponds to the dictionary definition of chaos, namely as 'utter confusion'. This is the meaning of chaos that we use in connection with the second law of thermodynamics, which we discussed earlier. We saw that processes occur in the direction of increasing entropy, which means increasing disorder, and we referred to an 'arrow of time' corresponding to increasing disorder. This means that the universe is becoming more and more disordered, so that eventually there will be complete chaos. Of course, it will be many billions of years before anything like that will happen. As we saw in the last chapter, even our Sun may survive in its present form for another several billion years, and long afterwards much will still be going on in the universe.

The type of chaos that is involved in the modern theory of chaos is much more limited. The word was first used in this special sense in 1975 by the mathematician James Yorke, and regrettably it proved popular with others working in the field and now we cannot escape from it. Chaos to its initiates relates to certain occurrences for which the final outcome is impossible to predict. The basic idea behind this limited type of chaos is that as time goes on the behaviour of some systems becomes progressively more sensitive to the initial conditions. This kind of chaos is obviously different from the wild and confused chaos that will exist at the end of time, and it is unfortunate that the same word has come to be used for it. Chaos in this much more limited sense is sometimes referred to as non-catastrophic chaos. It is also known as deterministic chaos, for reasons to be discussed.

One reason that this special use of the word chaos is regrettable is that it forces people to use expressions that seem rather comical. It is sometimes stated, for example, that the Solar System is 'slightly chaotic', by which is meant that the chaotic behaviour is hardly detectable at the present time, but might become evident in a million years or so. Saying 'slightly chaotic' sounds to me as ludicrous as saying that something is slightly unique, or that someone is slightly demented, slightly destitute—or slightly dead.

The following example may help us to understand what chaos is in this more limited sense. Imagine a lawn that becomes infested with beetles, and to keep things simple suppose that we do nothing about it, merely watching in dismay. The lawn deteriorates, and the beetle colony grows exponentially. The lawn becomes so bad that beetles die of starvation, and only a few remain. The lawn then starts to revive and becomes more respectable; the few beetles still alive begin to replicate, and the cycle starts over again. It is easy to see that the quality of the lawn will go through maxima and minima, the population of beetles doing the same.

What will be the final outcome? One possibility is that the grass and the beetles might all be dead. Another is that the lawn might settle down into an equilibrium state in which the beetle population remains fairly low and the lawn is not too bad but not at its best. Which would happen would be impossible to predict, even if there were no outside influences. This is true even if we ignore environmental factors such as the weather or the addition of chemicals, which would further influence the outcome. This is an example of what is called *predator–prey* chaos. It is also found with a population of cheetahs and antelopes, the cheetahs relying for their survival on a supply of antelopes as food, and the antelopes running away from them as fast as possible. The population of antelopes is then apt to fluctuate in a chaotic manner, and the population of the cheetahs will also fluctuate unpredictably.

Here is another example taken from the field of population dynamics. Suppose that we introduce a small number of fish into a lake that is completely free from external pollution. At first they will thrive in the favourable environment and reproduce rapidly. Soon, however, they will have to compete with one another for food, and will also suffer from the pollution that they themselves will create. Their population may then decline, and later recover. Under some circumstances the result may be a population that rises and falls in a completely unpredictable way.

The following familiar example brings about rather clearly why this type of chaos occurs. It applies to three well-known games: croquet, billiards, and curling. In each case one aspect of the game is that you propel a ball (or stone in the case of curling) in such a way that it first hits one ball, bounces off it, and then hits another. In billiards this is called making a cannon. It is relatively easy to hit the first ball, but it takes much more skill to propel the ball so precisely that it also hits the second one. If hypothetically you were called on to hit successively a whole series of balls it would become increasingly difficult. Only the most skilful of players would have much chance of hitting a third ball, and hitting the fourth and subsequent balls would be entirely a matter of luck. It is easy to see from this example what the essential feature of chaos is. We are dealing with a series of events, each one of which depends much more critically than the previous one on the precise conditions.

The following rather hypothetical argument brings out the point rather clearly. Suppose that you were at a billiards table and wanted to make a cannon, and that

I were standing watching you. Would you make a suitable adjustment for the gravitational force that I was exerting, taking into account just where I was standing? Of course not, particularly since (as we saw at the beginning of the last chapter) the gravitational force is such a fantastically weak force. Suppose on the other hand that you were ambitious enough to try to hit nine balls in succession; should you then take my gravity into account? The answer to this, according to the French mathematician Ivar Ekeland's fascinating book *Mathematics and the Unexpected* (1988), is that yes, you should. Of course, in practice it is quite impossible to do so, since the gravitational force is so tiny. But in principle my gravitational force does affect the outcome of a nine-ball cannon, because of the multiplication of the effects as successive collisions take place. Even the gravitational force of an electron at the far outer edge of the universe, 10^{10} light years away, will make a difference after a cannon of enough balls—after about 56 according to the calculations.

A pinball machine (Fig. 27) is also instructive in showing us how probability applies to chaotic motion. We assume that it is perfectly constructed so that if the ball is released just above the top pin, it next hits either of the pins on the second row with equal probability. If it hits the left pin of the second row it then has an

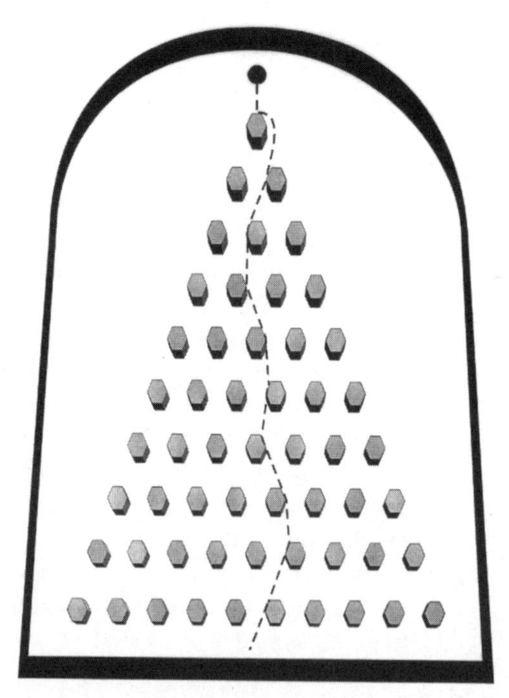

Fig. 27 A pinball machine. Since the path of the ball is very sensitive to its position, it is impossible to predict the outcome from the initial conditions, that is, the behaviour is chaotic. Where the ball finally lands, however, is not random since its probability can be calculated.

equal chance of hitting the first two pins of the third row. There are 10 rows in the machine shown in Fig. 27, and it is easy to calculate that there are exactly 512 ($= 2^9$) trajectories from the top to the bottom row. Only one of these leads to the left-hand pin in the bottom row, so that we know that the probability of hitting that pin is 1 in 512. The probability of hitting the right-hand pin in that row is also 1 in 512. There are higher probabilities of hitting the inner pins, and these can easily be worked out. To understand this more easily, consider just the fourth row. There is just one chance of hitting the left-hand and right-hand pins in this row. As to the other two, the second and third, it is easy to see that there are three trajectories leading to them. There are eight (2^3) trajectories in all, so that the chance of hitting the left-hand and right-hand pins is one in eight, while the chance of hitting the second or third pins is three in eight.

Note that the chance of hitting any one of the pins is determined by simple probability principles. There is an important distinction between chaos and randomness. Consider, for example, the bottom row of pins. If these pins were hit completely randomly, there would be an equal probability of hitting each one of them, instead of a much lower probability of hitting the outer ones as compared with the inner ones. Thus, although the motion of the ball is chaotic as it rolls down the machine, the probability that it hits a certain pin is determined by the construction of the pinball machine. We can express this by saying that we have *deterministic chaos*. This is the basis of the methods used to determine whether behaviour is random or chaotic. We apply probability tests to the data and if probability relationships apply we know that we have chaos; otherwise we have randomness.

Consider another familiar process, the bouncing of a tennis ball from a racquet. With practice we can hold a racquet horizontally and control the up and down motion of the bouncing ball in such a way that the ball always rises to the same height. Suppose, on the other hand, that we hit the ball with a predetermined force and a frequency that is independent of the motion of the ball. The ball will then bounce up to a different height each time. After a few bounces the height of a bounce will be quite unpredictable, even if we measure the conditions as precisely as possible. This is chaos, in the modern scientific sense of the word.

There are many other familiar examples of this kind of chaos. If we are pushing a child on a swing, we normally avoid the possibility of chaos by coordinating our pushing to the natural motion of the swing. If instead we perversely pushed at a fixed frequency, and without such coordination, the swing would perform a chaotic motion (anyone trying the experiment should not have a child on the swing). Our lives frequently involve avoiding chaos of this kind. If we drive off in a car and steer and brake in a predetermined manner, without regard to pedestrians or traffic, we will soon achieve chaos—and probably a court appearance.

Non-catastrophic chaos is often the outcome when there is oscillatory behaviour. An example is provided by heartbeat, which is controlled by natural pacemakers. Sometimes these do not work together properly, so that there are alternate long and short gaps between the beats. Under more extreme conditions the beating becomes highly irregular. In one particularly dangerous condition called *ventricular fibrillation* the heart flutters erratically instead of expanding and contracting rhythmically when it pumps the blood. A small change in the timing of one beat makes a bigger change in the timing of the next; the beating becomes chaotic. One feature of ventricular fibrillation that makes it difficult to treat is that the individual processes in the heart may be working normally; it is their lack of coordination that leads to overall malfunctioning.

There are, of course, many familiar processes which do not give rise to deterministic chaos. Examples are the swinging of a simple pendulum, ordinary engines such as those in automobiles, and most chemical reactions including those that occur in the body. For these we can set up and solve the appropriate mathematical equations; then, knowing the initial conditions, we can predict exactly and without any ambiguity just how the process will occur over a period of time. For such systems the behaviour may be complex and difficult to work out, and we may have to use numerical methods and use a computer. Often the results present no surprises: what occurs is just what we expect. This is fortunate in many practical situations; it would be disastrous if, when we applied the brakes to a car, it would sometimes accelerate for no ascertainable reason.

It is rather a paradox that although chaos theory has developed so recently it deals with many things with which we are all familiar: water running from a tap, a flag flapping in the wind, traffic jams on the roads, the rise and fall of the stock market, and the rise and fall of civilizations. It deals with such common occurrences as a political protest which may start with everyone behaving in a reasonable manner, but which ends in a violent riot. The theory turns everyday happenings, previously avoided and sometimes scorned by scientists, into legitimate subjects of scientific study. Chaos theory unites research workers in different fields and works against the excessive specialization that is so prevalent today.

Up until the nineteenth century the prevailing scientific theories were *deterministic*, by which we mean that the future is entirely determined by the past. This had been assumed to be true by Isaac Newton, and a century later Pierre Simon Laplace (1749–1827; Fig. 28) argued that if we could specify the exact state of the universe at a given time we could in principle deduce its state at any future time. Laplace was born in Normandy and attended the University of Caen where he was originally intended to be a cleric. His mathematical gifts were soon recognized, and he was appointed professor of mathematics at the École Militaire. He was elected a member of the Académie des Sciences at the early age of 24. He was one of the most versatile and influential scientists of all time. He made valuable contributions to a wide variety of problems in mathematics and physics, he developed a highly productive philosophy of science, and he played an important part in establishing the modern scientific disciplines. He was interested in experimental science as well as in pure and applied mathematics.

From 1768 Laplace worked on the integral calculus, astronomy, cosmology, the theory of games, probability, and causality. During this period he entered into a

Fig. 28 Pierre Simon Laplace, Marquis de Laplace (1749–1827), distinguished for his work in pure mathematics, physics, and astronomy. He made important contributions to probability theory, and suggested a comprehensive theory of the universe.

collaboration with the great French chemist Antoine Lavoisier (1743–1794) on theoretical and experimental investigations of heat. During the revolutionary period in France, from 1789 to 1805, Laplace's reputation was at its height. Always a wily and calculating opportunist, he shifted his political opinions as occasion demanded, and always managed to retain a position of influence. He was an active member of the commission that introduced the metric system in 1799, and he played a dominant role in the creation of the Institut de France and the École Polytechnique. The first four volumes of his comprehensive *Traité de méchanique céleste* (Treatise on celestial mechanics) appeared from 1799 to 1805, with later instalments appearing from 1823 to 1825; it was these later volumes that contained much of Laplace's physics. Laplace presented a copy of the book to Emperor Napoleon I who commented that, although dealing with the universe, the book made no mention of its Creator. To this Laplace stiffly and bluntly replied: '*Je n'avais pas besoin de cette hypothèse-là*' (I had no need for that hypothesis). Napoleon, much amused, told the story to many people, including the distinguished mathematician Joseph Louis Lagrange (1736–1813) whose comment was: '*Ah! c'est une belle hypothèse; ça explique beaucoup de choses!*' (Ah! that is a nice hypothesis; it explains many things!).

For a short time Laplace served as minister of the interior under Napoleon, but he was ineffective as an administrator; Napoleon commented later that Laplace 'saw no question from its true point of view; he saw subtleties everywhere, had only doubtful ideas, and finally carried the spirit of the infinitely small into the management of ideas.' After six weeks as minister Laplace was dismissed by Napoleon, who nevertheless continued to hold him in high esteem and later appointed him to the Senate. When in 1814 the monarchy was restored in France, Laplace hastened to offer his services to the Bourbons, and was rewarded with the title of marquis. Like his contemporary Count Rumford, whom we met in Chapter 1, he was vain and self-seeking in some matters, but could be generous and considerate under other circumstances.

We saw in Chapters 5 and 6 that the first important challenge to determinism came from the interpretation of the second law of thermodynamics in terms of probability. That interpretation showed that events do not always follow from one another by strictly deterministic laws, but sometimes occur as a matter of pure chance. Another kind of challenge came at about the same time from the French mathematician, physicist, and philosopher Jules Henri Poincaré (1854–1912; Fig. 29). He was born in Nancy, France, the son of a physician, and educated at the École Polytechnique and the École des Mines in Paris. He was soon recognized as a mathematician of great brilliance and originality, and at the early age of 27 was appointed professor of mathematics at the Université de Paris, where he remained for the rest of his life. He soon gained a reputation as the last of the great universalists, since he contributed to a vast range of topics in mathematics and applied science. In spite of this he was singularly absent-minded and clumsy, being

Fig. 29 Jules Henri Poincaré (1854–1912), who worked in a wide range of mathematical problems. Because of the breadth of his interests he has often been referred to as the last of the universalists.

quite inept with the simplest of mathematics including arithmetic. He was not elected to the exclusive Académie des Sciences until he was 55, which is surprising in view of his great distinction. There is a story that before he was elected to that body a friend asked him if he minded not having been made a member. He made the philosophical reply that he preferred to be a non-member with people wondering why he had not been elected, than to be a member with people wondering why he had been elected.

His challenge to determinism came as a result of his mathematical attack on what is called the three-body problem. An example of this is the Sun–Earth–Moon system, which Newton had attacked two centuries earlier. Newton had successfully solved the dynamics of a planet such as the Earth moving in its orbit round the Sun. This was one of his most important contributions to what Laplace later called celestial mechanics, and he then attacked the problem of the Earth moving round the Sun with in addition the Moon moving in its orbit round the Earth. At first sight one might think that that such a three-body system would be solvable by use of the same procedures, but this is by no means the case. Newton could not find an exact solution to the dynamical equations, in the form of equations that described the motions of the three bodies, for the simple reason that there is no such solution.

Having failed to find what is called an explicit solution to the problem, Newton resorted to a technique mathematicians refer to as the 'perturbation method', which provides an approximate solution. The principle involved is to start with the solution for the Sun–Earth problem, which is the dominant problem. The influence of the Moon he then added on as a perturbation to this solution. This proved exceedingly difficult and he worked on it intensely for a full year. He finally obtained a solution, but was disappointed to find that it did not give the orbit of the Moon as accurately as he had wished. Newton always considered his work on the Moon to be his great failure, later commenting 'Never did I get bigger headaches than when I was working on the problem of the Moon.' This failure may well have contributed to the fact that in 1696 he resigned his professorship at Cambridge University and spent most of the rest of his life in senior administrative positions at the Royal Mint, supervising the minting of new currency and relentlessly prosecuting criminals who were involved in offences such as forgery and clipping coins for the precious metals they contained.

Subsequent to Newton's work on the three-body problem several eminent mathematicians also tackled it, but with little success. They all used the perturbation methods, and could not overcome the fundamental difficulty that the perturbation introduced by the Moon was just too big to lead to satisfactory results. Poincaré's success came as a result of introducing a completely new method, one that is used to this day. Instead of using the ordinary three-dimensional space with which we are familiar he used *phase space*. The basic idea of this is as follows. The conventional three dimensions are dimensions of position only, but we can think of another three dimensions relating to velocity. The motion of a ball thrown into the air can, for example, be described by three position dimensions and three dimensions of velocity, by which we mean the components of the velocity along three axes.

By working in this way on the three-body problem Poincaré was led to completely new insights. He began the work as a result of a mathematics contest organized in 1889 by the University of Stockholm to mark the 60th birthday of Oscar II, King of Sweden and Norway (which were then one country). The problem set in the contest was to determine the stability of the Solar System. Will the planets remain in their precise orbits for ever, always retracing the same paths, in which case the Solar System is stable? Alternatively, will the orbits of the planets, as a result of the cumulative effects of the gravitational attractions of other planets, change radically in the distant future? Previous mathematical treatments had led to opposite conclusions, and the purpose of the contest was to come to a satisfying decision.

Having introduced the phase-space technique Poincaré, aged 35 at the time, felt in a good position to enter the Stockholm contest. As he worked on the problem he was led to the conclusion that in certain complex dynamical situations such as the three-body problem a minute change in the initial position or velocity of one of the three bodies could lead to a complete change in its orbit. A tiny change, in other words, could lead from order to chaos—in its limited sense of unpredictability, of course. Regularity and chaos, he found, are not completely separate situations but are closely interrelated, the unpredictable being never far away from the predictable. What seems a fairly simple system such as the Sun–Earth–Moon system, although precisely governed by Newton's gravitational laws, can give rise to unpredictable behaviour. The deterministic universe of Newton and Laplace is therefore not a valid one.

Poincaré won the contest of the University of Sweden, and in 1889 his 270-page article, entitled 'Sur le problème des trois corps et les équations de la dynamique', was printed in an issue of the *Acta Mathematica*, an increasingly influential journal. Soon after the issue appeared one of its readers pointed out a significant error in Poincaré's treatment. This gave rise to serious complaints that another contestant, who had submitted a satisfactory treatment of the problem but not as thorough as that of Poincaré, should have been the winner. The editor of the *Acta* was the chief organizer of the contest, the mathematician Gösta Mittag-Leffler (1846–1927), and he was naturally disturbed by this controversy, feeling that it would sully the reputations of those who had been concerned with selecting the winner of the contest—why had not they detected the error in the Poincaré submission? Mittag-Leffler took rather a drastic action: instead of publishing a correction in a subsequent issue he diligently tracked down to their destinations as many of the offending issues as he could, and destroyed them. He did this so successfully that today only one of the issues is known to exist, and it is kept inaccessible under lock and key in the Mittag-Leffler Institute near Stockholm.

At the same time Mittag-Leffler persuaded Poincaré to correct his error and to submit a revised paper which would be published in a subsequent issue of the *Acta* as if it had been the prize-winning paper. Poincaré never denied that he had made an error, although in view of the secrecy involved no one knows just what it was. It presumably was not of vital significance as Poincaré soon produced a satisfactory proof. Between the award ceremony and the appearance of the revised paper in 1890 there was a good deal of behind-the-scenes bickering, but it soon died down and in the end Mittag-Leffler's cover-up had achieved its purpose.

In Poincaré's later book *Science et Méthode* (1909) he summarized his important conclusion as follows:

> There are situations where small differences in the initial conditions can produce very large ones in the final result: a small error in the former can lead to a huge error in the latter. In those cases, predictions become impossible.

This is a very clear statement of the basic idea of modern chaos theory, but few people at the time paid any attention to it, and the idea was not revived until the latter half of the next century.

Another kind of challenge to determinism came in the 1920s with the advent of quantum mechanics and in particular with Heisenberg's formulation of the uncertainty principle, which we discussed in Chapter 7. According to this principle it is impossible to determine, at one and the same time, the exact position and momentum of a particle. As a result, we cannot predict the exact course of events, but can sometimes make estimates of probabilities.

Non-catastrophic chaos theory, however, relates to a *deterministic* type of uncertainty, which would occur even if there were no restriction imposed by the Heisenberg principle. Even if the uncertainty principle did not affect the situation we still could not always predict what is going to occur. The reason appears when one analyses the situation mathematically. With certain kinds of dynamical equations, and under certain other conditions which we will look at later, there is a basic instability in the behaviour.

Poincaré's conclusions were largely ignored for more than half a century. Revival of interest in them was brought about to a considerable extent by the development of computers, which enabled numerical calculations to be made much more rapidly than before. The laborious calculations on the Sun–Earth–Moon system that Newton carried out over a period of about a year can be made today in a few seconds, so that much more extensive calculations can now be made. It has been commented that that no great scientific discovery has ever been made by means of computers, and I think this is still true. Computers nevertheless made an important contribution in consolidating the conclusion that Poincaré had reached about the unpredictability of the behaviour of certain systems.

One important example of this is to be found in some computer investigations carried out in the1960s by the American meteorologist Edward Lorenz (b. 1917) of the Massachusetts Institute of Technology. In order to study weather patterns he used a computer, and devised highly simplified differential equations to represent meteorological processes such as the movement of air and the evaporation of water. His aim was to make predictions of the temperature, the direction of the wind, and the onset of rain or snow at later times. He introduced into his programs absolutely precise initial conditions and then let the computer run in order to predict the values of the various meteorological parameters after various periods of time. To his great surprise he found, even with his highly simplified model, that when he repeated the calculations with essentially the same initial conditions the predictions might be much the same for the first few days, but after that there were enormous divergences, no similarity between the predictions remaining after more than a few days (Fig. 30). At first he assumed that the computer was malfunctioning, but much further work showed the effect to be real.

This means, of course, that there are fundamental limits to our ability to predict the weather. There are problems even when the mathematical equations have been greatly simplified, and they are much greater for the conditions actually existing. Experts can fairly reliably predict the weather for the next two or three days. After a few more days, however, computers will make a wide range of predictions from closely similar starting conditions, which means that predictions are in reality impossible. In 1972 Lorenz suggested what is called the 'butterfly effect', with his paper entitled 'Does the flap of a butterfly's wings in Brazil set off a tornado in Texas?' A comment from an expert on weather forecasting nicely sums up the situation: 'Today we can predict the weather very precisely, provided that it does not change unexpectedly.'

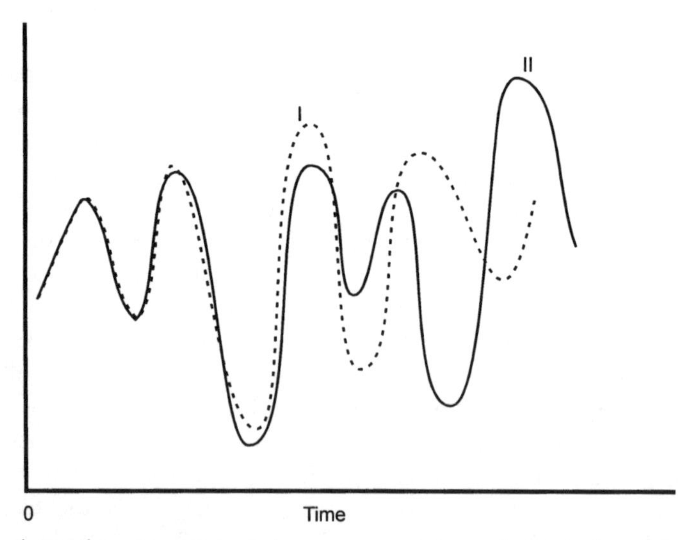

Fig. 30 A weather parameter, such as temperature or wind velocity, plotted against time, as predicted by a computer for two identical initial conditions (curves I and II). Edward Lorenz found that the curves were more or less identical for the first two or three days, but that they later diverged more and more, so that it is impossible to make predictions for more than a week or so.

The conditions under which chaos can develop have been worked out. One is that at least three individual processes must be involved; this condition is usually satisfied. Another condition is that at least one of the processes must be non-linear in the mathematical sense; this means that the effect should not be directly proportional to the cause, but should depend upon it more strongly. Another condition is that there must be *feedback*.

We met feedback in the last chapter, where we saw that negative feedback occurs with the thermostats which control the temperature in our homes, and also controls the temperatures of stars. Positive feedback is illustrated by a thermostat which has been wired by an unreliable electrician so that if the temperature is too low the furnace goes off, and if it is too high the furnace goes on. If then the temperature happens to be too high, the furnace makes it still higher and the temperature eventually settles down at some high value, corresponding to the best that the furnace can do. Positive feedback happens in a chemical or nuclear explosion; a nuclear explosion, for example, occurs when uranium-235 reaches a critical mass.

A familiar example of positive feedback which many of us have experienced at first hand is at an event such as a cocktail party. What often happens is that the sound level soon becomes so high that in order to communicate we have to shout at the top of our voices (the speed with which this state is attained is roughly proportional to the amount and availability of alcohol). It would obviously make more sense if everyone agreed to speak in a normal voice, but that would only work if police were assigned to haul away offenders, and that is probably not the most popular way to give a party. If there were such a prior agreement but no police we can be sure that someone would soon speak a little louder, forcing others to do the same, and the sound would soon reach a high level.

In all the examples of chaos we have considered it is easy to see that there is feedback. In the case of the beetles affecting the growth of grass, the number of beetles affects (adversely) the growth of the grass, while the amount of grass affects (favourably) the growth of the beetle colony; we have negative feedback in one direction and positive in the other.

A simple mathematical equation is helpful in the understanding of chaos. Suppose we are concerned with some property, such as the population of fish in a pond, which we will represent by the symbol x. The particular value that the population has at a time t we will denote as x_t; that after a certain fixed interval of time (e.g. a day, a week, or a year) we denote as x_{t+1}. If the interval is one day, the time t is the number of days after the fish were put in the lake; that day would be day zero, the next day would be day one, and so on. If the population were stable, neither increasing nor decreasing, the population on any day would be the same as that on the previous day, and we could express this by saying that $x_{t+1} = x_t$. Suppose on the other hand that the fish were reproducing so that the population on any day was greater that that on the previous day. A simple way of expressing this is to write the equation

$$x_{t+1} = \lambda x_t$$

where λ (Greek lambda) is a number (not necessarily a whole number) which is greater than one. For example, if $\lambda = 2$, the population of fish is doubling every day; if it is 1.1 the population is increasing by 10 per cent every day, and so on.

This, of course, cannot go on indefinitely, because after a time the lake will not be able to produce enough food, and will become polluted. The rate of growth of the population will therefore eventually slow down. A simple way of representing this is to modify the last equation by subtracting a term λx_t^2:

$$x_{t+1} = \lambda x_t - \lambda x_t^2 = \lambda x_t(1 - x_t)$$

Here a little clarification is needed. Since the factor $(1 - x_t)$ enters into this equation it follows that x_t can never be greater than one, since otherwise the next x value would be negative, which of course is impossible. We must therefore regard x as the value (such as the population) relative to a certain population which is never exceeded.

Of course, we could more generally use for the second term a coefficient that is different from λ, but it is mathematically simpler to use the same value and this is what is often done. The term λx_t^2 is, of course, a feedback term; it corresponds to negative feedback since it corresponds to the fact that factors are having a deleterious effect on the rate of growth. We should note also that this last equation is a non-linear equation, since it involves x_t^2, the quantity x_t raised to a power other than one, namely the second power.

We need not go into details; it will be enough to consider the kinds of behaviour to which this last equation can lead. Sometimes they are unexpected—which is of course the essence of chaotic behaviour. Everything depends on the value of the coefficient λ. If it is between 1 and 3 the behaviour is quite straightforward; the population x finally settles down at a constant value, as shown in Fig. 31(a). If λ is between 3 and 3.57 the population first rises and then oscillates between two fixed values (Fig. 31(b)). If λ is greater than 3.57 there is again oscillation, but now between unpredictable values rather than the same two values (Fig. 31(c)). This case also has another characteristic, typical of chaos. The details of the oscillations now vary greatly with the initial conditions; a tiny change in them leads to completely different behaviour.

Chaos theory has important implications for our understanding of the evolution of the universe and of life in it. In the past there has been uneasiness over the fact that the formation of the galaxies, with their complicated stellar and planetary systems, seemed to imply the creation of order out of disorder, in violation of the second law of thermodynamics. The same difficulty seemed to arise for the formation of complex structures, such as the human eye, found in the higher life forms; how, it has been argued, can such complicated organization arise from much more disordered states? Snowflakes have beautifully complex structures; how can so many myriads of them descend upon us from highly disorganized storm clouds without there being a violation of the second law?

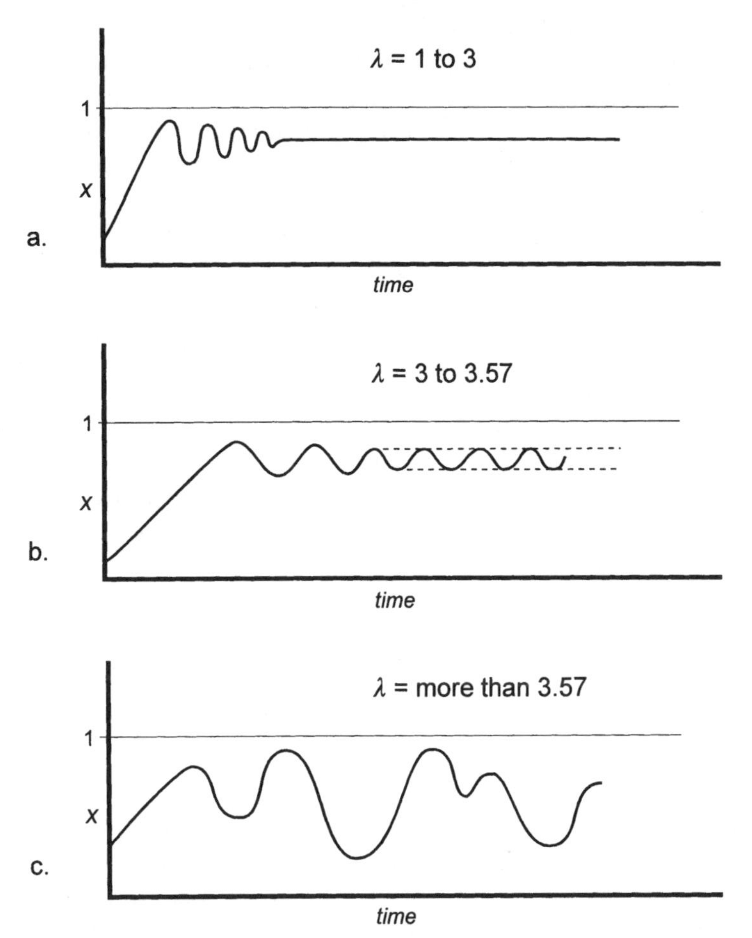

Fig. 31 The shapes of the x versus t curves arising from the equation $x_{t+1} = \lambda x_t - \lambda x_t^2 = \lambda x_t(1 - x_t)$ for various values of the coefficient λ. (a) λ is between 1 and 3. The quantity x rises, perhaps with oscillations, but finally settles down at a constant value. (b) λ is between 3 and 3.57. The quantity x finally oscillates between two values. (c) λ is greater than 3.57. The quantity x oscillates in an irregular manner, and the behaviour is highly sensitive to the initial conditions. This is chaos.

Chaos theory provides the answer. The second law applies only to the final state of a process; intermediate states can show oscillations, or complex types of behaviour such as are sometimes found in experiments on complex chemical reactions. Similarly, the mathematical physicist Benoit Mandelbrot (b. 1924) discovered that chaos theory leads to the possibility of fascinating complex patterns which he called *fractals*, of which a snowflake and a cauliflower are familiar examples. There is thus no theoretical difficulty about the formation of complex structures. The second law imposes the condition that there must be an increase in disorder in the overall process; the intermediate states can be ordered.

It is often stated that two scientific advances, the concepts of relativity and of quantization, were the two greatest scientific achievements of the twentieth century. A good case can be made for regarding chaos theory as of equal significance for our understanding of the universe around us.

The wise person makes predictions only with caution, recognizing that the best we can do is to express our opinion as to what is most probable. Public figures tend to predict the future with great confidence. Presidents, prime ministers, finance ministers, and governors of banks are apt to make flat statements that the economy, although shaky at the present time, will certainly improve after a short time. The public predictions of prominent people are, of course, more than pure predictions; they influence the future. A government leader who predicts that a country will collapse economically will almost certainly cause it to collapse. Leaders therefore tend to be optimistic, even against the weight of the evidence, knowing that it may help to make things better.

Similarly, a widely publicized prediction from a respected source that the stock market will fall will almost certainly make it fall. The stock market provides us with an excellent example of instability resulting from positive feedback. If the market begins to fall, as soon as the word gets out investors will sell, causing the market to fall further. There are some safeguards that have the reverse effect, as otherwise recovery could never occur. When prices become low enough investors will take advantage of the low prices and will buy. It is, however, a matter of concern that the stock market is so susceptible to the whims and sometimes panic of investors.

An uncertainty principle applies as much to human affairs as it does to the world of elementary particles. As we saw in Chapter 7, Werner Heisenberg recognized that there is a fundamental difficulty about making a simultaneous measurement of the position and speed of a particle; if we try to find out exactly where it is we are bound to disturb it, and then do not know just where it is or where it is going. This principle of uncertainty applies also to some who make public predictions; their predictions affect the future. The reason that public figures tend to make optimistic predictions is that they realize—perhaps instinctively rather than explicitly—that by making such predictions they are making it more likely that there will be a happy outcome.

Predictions in the fields of science and technology are no more reliable than those in the political and financial worlds. The accompanying table lists some of the predictions that experts have made in the past and which have been proved wrong by subsequent events. Much of the technology that we employ today was never thought of half a century ago. The lesson is that good scientific research should be carried out even if it does not appear to have any applications. Most of what is useful today was made possible by research that was done without any attempt to predict the future, a task we now know to be fruitless.

Some bad predictions

Statement	Source	Year
Invention		
Everything that can be invented has been invented	Charles H. Duell, Commissioner, US Patents	1899
Telephone		
This telephone is inherently of no value to us	Western Union internal memorandum	1876
It is certainly a wonderful instrument, although I suppose not likely to come to any practical use	Lord Rayleigh	1876
Radio and television		
Speech over the radio is as likely as a man jumping over the moon	Thomas Edison, said to R. A. Fessenden, the world's first disk jockey	1899
I don't think this business of television is likely to come to much	Sir J. J. Thomson	1930
Aircraft		
Heavier-than-air flying machines are impossible	Lord Kelvin	1885
I have not the smallest amount of faith in aerial navigation other than ballooning	Lord Rayleigh	1889
The gas turbine could hardly be considered a feasible application to airplanes	US National Academy of Sciences, Committee on GasTurbines	1940
Nuclear power		
Anyone who expects a source of power from the transformation of atoms is talking moonshine	Lord Rutherford (his opinion had changed a year or two later)	1933
Computers		
I think that there is a world market for maybe five computers	Thomas Watson	1943
There is no reason anyone would want a computer in their home	Ken Olson, founder of Digital Equipment Corporation; chairman of IBM	1977
640K ought to be enough for anybody	Bill Gates	1981
Space travel		
The possibility of travel in space seems at present to appeal to schoolboys rather than scientists	Sir George P. Thomson	1956
Space travel is utter bilge	Sir Richard Woolley, Astronomer Royal	1956
Antibiotics		
It is time to close the book on infectious disease	US Surgeon-General	1969

Suggested reading

There are many books dealing with aspects of the subject matter of the present book. Here I have listed a small selection of books that I think will help the reader to delve more deeply into some of the topics.

Brown, G. I. (1999). *Count Rumford: The Extraordinary Life of a Scientific Genius.* Sutton, Stroud.

Campbell, John (1999). *Rutherford: Scientist Supreme.* AAS Publications, Christchurch, New Zealand.

Cardwell, D. S. L. (1971). *From Watt to Clausius.* Manchester University Press and Cornell University Press.

Davies, Paul (1999). *The Fifth Miracle: The Search for the Origin and Meaning of Life.* Simon and Schuster, New York.

Ekeland, Ivar (1988). *Mathematics and the Unexpected.* University of Chicago Press. On p. 68 of this book is quoted the reference for the effect of gravity on a billiard cannon: M. Berry, 'Regular and irregular motion', in *Topics in Non-Linear Dynamics*, American Institute of Physics Conference Proceedings, No. 46, American Institute of Physics, pp. 111–112, 1978.

Gamow, George. *Thirty Years that Shook Physics: The Story of Quantum Theory.* Doubleday, New York, 1966; Dover Publications, New York, 1985.

Gleick, James (1987). *Chaos: Making a New Science.* Penguin Books, Harmondsworth.

Gribbin, John (1998). *Almost Everyone's Guide to Science: The Universe, Life, and Everything.* Yale University Press.

Gribbin, John (1998). *In Search of the Big Bang: The Life and Death of the Universe.* Penguin Books, Harmondsworth.

Hall, A. Rupert (1992). *Isaac Newton: Adventurer in Thought.* Cambridge University Press.

Hall, Nina (ed.) (1992). *The New Scientist Guide to Chaos.* Penguin Books, Harmondsworth.

Laidler, Keith J. (1998). *To Light Such a Candle.* Oxford University Press.

Leff, Henry S. and Rex, Andrew F. (1990). *Maxwell's Demon.* Princeton University Press.

Lindley, David (2001). *Boltzmann's Atom: The Great Debate that Launched a Revolution in Physics.* The Free Press, New York.

Rees, Sir Martin (1997). *Before the Beginning.* Addison-Wesley, New York.

Reeves, Hubert (1993). *Atoms of Silence.* Stoddart, Toronto.

Smith, C. W. and Wise, M. N. (1989). *Energy and Empire: A Biographical Study of Lord Kelvin.* Cambridge University Press.

Stewart, Ian (1990). *Does God Play Dice? The New Mathematics of Chaos.* Penguin, Harmondsworth.

Thomas, Sir John Meurig (2001). 'Predictions'. *Notes and Records of the Royal Society*, Vol. 55, pp. 105–117.

Thuan, Trinh Xuan (2001). *Chaos and Harmony: Perspectives on Scientific Revolutions of the Twentieth Century.* Oxford University Press, New York.

von Baeyer, Hans Christian (1998). *Maxwell's Demon.* Random House, New York. Paperback edition entitled *Warmth Disperses and Time Passes: The History of Heat.* Modern Library, New York, 1999.

Westfall, Richard S. (1980). *Never at Rest: A Biography of Isaac Newton.* Cambridge University Press.

White, Michael and Gribbin, John (1993). *Einstein: A Life in Science.* Simon and Schuster, London.

Index